ISEE Upper Level
Math Tutor

Everything You Need to Help Achieve

an Excellent Score

By

Reza Nazari

&

Ava Ross

EffortlessMath

About Effortless Math Education

Effortless Math Education operates the www.effortlessmath.com website, which prepares and publishes Test prep and Mathematics learning resources. Effortless Math authors' team strives to prepare and publish the best quality Mathematics learning resources to make learning Math easier for all. We Help Students Learn to Love Mathematics.

All inquiries should be addressed to:
info@effortlessMath.com
www.EffortlessMath.com

ISBN: 978-1-64612-847-1

Published by: **Effortless Math Education**

for Online Math Practice Visit www.EffortlessMath.com

ISEE Upper Level MATH TUTOR

All the Tools You Need to Succeed on the ISEE Upper Level Math test 2020!

Feeling anxious about the ISEE Upper Level? Not sure your math skills are up to the task? Don't worry, *ISEE Upper Level Math Tutor* has you covered!

Focusing on proven test-taking strategies, easy-to-understand math principles, and professional guidance, *ISEE Upper Level Math Tutor* is your comprehensive study guide for the ISEE Upper Level Math test!

Each chapter includes a study-guide formatted review and quizzes to check your comprehension on the topics covered. With this self-study guide, it's like having your own tutor for a fraction of the cost!

What does the ISEE Upper Level Math Tutor offer?

- Content 100% aligned with the 2020 ISEE Upper Level test

- **Step-by-Step guides** to all ISEE Upper Level Math concepts and topics covered in the 2020 test

- **Over 500 additional ISEE Upper Level math practice questions** featuring multiple-choice and grid-in formats with answers grouped by topic, so you can focus on your weak areas

- Abundant Math skill-building exercises to help test-takers approach different question types that might be unfamiliar to them

- **2 full-length practice tests** (featuring new question types) with detailed answers.

The surest way to succeed on the ISEE Upper Level Math Test is with intensive practice in every math topic tested—and that's what exactly what you'll get! With the ISEE Upper Level Math Tutor, you'll have everything you need to ace the ISEE Upper Level right in your hands. *Start studying today!*

About the Author

Reza Nazari is the author of more than 100 Math learning books including:

- ❖ **Math and Critical Thinking Challenges:** For the Middle and High School Student
- ❖ **ACT Math in 30 Days**
- ❖ **ASVAB Math Workbook 2018 - 2019**
- ❖ **Effortless Math Education Workbooks**
- ❖ **and many more Mathematics books**

Reza is also an experienced Math instructor and a test–prep expert who has been tutoring students since 2008. Reza is the founder of Effortless Math Education, a tutoring company that has helped many students raise their standardized test scores—and attend the colleges of their dreams. Reza provides an individualized custom learning plan and the personalized attention that makes a difference in how students view math.

You can contact Reza via email at:
reza@EffortlessMath.com

Find Reza's professional profile at:
goo.gl/zoC9rJ

Contents

CHAPTER 1:

FRACTIONS AND MIXED NUMBERS

Math Topics that you'll learn in this chapter:

▶ Simplifying Fractions

▶ Adding and Subtracting Fractions

▶ Multiplying and Dividing Fractions

▶ Adding Mixed Numbers

▶ Subtracting Mixed Numbers

▶ Multiplying Mixed Numbers

▶ Dividing Mixed Numbers

SIMPLIFYING FRACTIONS

☑ A fraction contains two numbers separated by a bar between them. The bottom number, called the denominator, is the total number of equally divided portions in one whole. The top number, called the numerator, is how many portions you have. And the bar represents the operation of division.

☑ Simplifying a fraction means reducing it to the lowest terms. To simplify a fraction, evenly divide both the top and bottom of the fraction by $2, 3, 5, 7$, etc.

☑ Continue until you can't go any further.

Examples:

Example 1. Simplify $\frac{12}{30}$

Solution: To simplify $\frac{12}{30}$, find a number that both 12 and 30 are divisible by. Both are divisible by 6. Then: $\frac{12}{30} = \frac{12 \div 6}{30 \div 6} = \frac{2}{5}$

Example 2. Simplify $\frac{64}{80}$

Solution: To simplify $\frac{64}{80}$, find a number that both 64 and 80 are divisible by. Both are divisible by 8 and 16. Then: $\frac{64}{80} = \frac{64 \div 8}{80 \div 8} = \frac{8}{10}$, 8 and 10 are divisible by 2, then: $\frac{8}{10} = \frac{4}{5}$ or $\frac{64}{80} = \frac{64 \div 16}{80 \div 16} = \frac{4}{5}$

Example 3. Simplify $\frac{20}{60}$

Solution: To simplify $\frac{20}{60}$, find a number that both 20 and 60 are divisible by. Both are divisible by 20, then: $\frac{20}{60} = \frac{20 \div 20}{60 \div 20} = \frac{1}{3}$

ADDING AND SUBTRACTING FRACTIONS

☑ For "like" fractions (fractions with the same denominator), add or subtract the numerators (top numbers) and write the answer over the common denominator (bottom numbers).

☑ Adding and Subtracting fractions with the same denominator:

$$\frac{a}{b} + \frac{c}{b} = \frac{a+c}{b} \qquad \frac{a}{b} - \frac{c}{b} = \frac{a-c}{b}$$

☑ Find equivalent fractions with the same denominator before you can add or subtract fractions with different denominators.

☑ Adding and Subtracting fractions with different denominators:

$$\frac{a}{b} + \frac{c}{d} = \frac{ad+bc}{bd} \qquad \frac{a}{b} - \frac{c}{d} = \frac{ad-bc}{bd}$$

Examples:

Example 1. Find the sum. $\frac{3}{4} + \frac{1}{3} =$

Solution: These two fractions are "unlike" fractions. (they have different denominators). Use this formula: $\frac{a}{b} + \frac{c}{d} = \frac{ad+cb}{bd}$

Then: $\frac{3}{4} + \frac{1}{3} = \frac{(3)(3)+(4)(1)}{4 \times 3} = \frac{9+4}{12} = \frac{13}{12}$

Example 2. Find the difference. $\frac{4}{5} - \frac{3}{7} =$

Solution: For "unlike" fractions, find equivalent fractions with the same denominator before you can add or subtract fractions with different denominators. Use this formula:

$\frac{a}{b} - \frac{c}{d} = \frac{ad-bc}{bd}$

$\frac{4}{5} - \frac{3}{7} = \frac{(4)(7)-(3)(5)}{5 \times 7} = \frac{28-15}{35} = \frac{13}{35}$

MULTIPLYING AND DIVIDING FRACTIONS

☑ Multiplying fractions: multiply the top numbers and multiply the bottom numbers. Simplify if necessary. $\frac{a}{b} \times \frac{c}{d} = \frac{a \times c}{b \times d}$

☑ Dividing fractions: Keep, Change, Flip

☑ Keep the first fraction, change the division sign to multiplication, and flip the numerator and denominator of the second fraction. Then, solve!

$$\frac{a}{b} \div \frac{c}{d} = \frac{a}{b} \times \frac{d}{c} = \frac{a \times d}{b \times c}$$

Examples:

Example 1. Multiply. $\frac{5}{8} \times \frac{2}{3} =$

Solution: Multiply the top numbers and multiply the bottom numbers.
$\frac{5}{8} \times \frac{2}{3} = \frac{5 \times 2}{8 \times 3} = \frac{10}{24}$, simplify: $\frac{10}{24} = \frac{10 \div 2}{24 \div 2} = \frac{5}{12}$

Example 2. Solve. $\frac{1}{3} \div \frac{4}{7} =$

Solution: Keep the first fraction, change the division sign to multiplication, and flip the numerator and denominator of the second fraction.
Then: $\frac{1}{3} \div \frac{4}{7} = \frac{1}{3} \times \frac{7}{4} = \frac{1 \times 7}{3 \times 4} = \frac{7}{12}$

Example 3. Calculate. $\frac{3}{5} \times \frac{2}{3} =$

Solution: Multiply the top numbers and multiply the bottom numbers.
$\frac{3}{5} \times \frac{2}{3} = \frac{3 \times 2}{5 \times 3} = \frac{6}{15}$, simplify: $\frac{6}{15} = \frac{6 \div 3}{15 \div 3} = \frac{2}{5}$

Example 4. Solve. $\frac{1}{4} \div \frac{5}{6} =$

Solution: Keep the first fraction, change the division sign to multiplication, and flip the numerator and denominator of the second fraction.
Then: $\frac{1}{4} \div \frac{5}{6} = \frac{1}{4} \times \frac{6}{5} = \frac{1 \times 6}{4 \times 5} = \frac{6}{20}$, simplify: $\frac{6}{20} = \frac{6 \div 2}{20 \div 2} = \frac{3}{10}$

ADDING MIXED NUMBERS

Use the following steps for adding mixed numbers:

☑ Add whole numbers of the mixed numbers.

☑ Add the fractions of the mixed numbers.

☑ Find the Least Common Denominator (LCD) if necessary.

☑ Add whole numbers and fractions.

☑ Write your answer in lowest terms.

Examples:

Example 1. Add mixed numbers. $3\frac{1}{3} + 1\frac{4}{5} =$

Solution: Let's rewriting our equation with parts separated, $3\frac{1}{3} + 1\frac{4}{5} = 3 + \frac{1}{3} + 1 + \frac{4}{5}$. Now, add whole number parts: $3 + 1 = 4$

Add the fraction parts $\frac{1}{3} + \frac{4}{5}$. Rewrite to solve with the equivalent fractions.

$\frac{1}{3} + \frac{4}{5} = \frac{5}{15} + \frac{12}{15} = \frac{17}{15}$. The answer is an improper fraction (numerator is bigger than denominator). Convert the improper fraction into a mixed number: $\frac{17}{15} = 1\frac{2}{15}$. Now, combine the whole and fraction parts: $4 + 1\frac{2}{15} = 5\frac{2}{15}$

Example 2. Find the sum. $1\frac{2}{5} + 2\frac{1}{2} =$

Solution: Rewriting our equation with parts separated, $1 + \frac{2}{5} + 2 + \frac{1}{2}$. Add the whole number parts:

$1 + 2 = 3$. Add the fraction parts: $\frac{2}{5} + \frac{1}{2} = \frac{4}{10} + \frac{5}{10} = \frac{9}{10}$

Now, combine the whole and fraction parts: $3 + \frac{9}{10} = 3\frac{9}{10}$

SUBTRACT MIXED NUMBERS

Use these steps for subtracting mixed numbers.

☑ Convert mixed numbers into improper fractions. $a\frac{c}{b} = \frac{ab+c}{b}$

☑ Find equivalent fractions with the same denominator for unlike fractions. (fractions with different denominators)

☑ Subtract the second fraction from the first one. $\frac{a}{b} - \frac{c}{d} = \frac{ad-bc}{bd}$

☑ Write your answer in lowest terms.

☑ If the answer is an improper fraction, convert it into a mixed number.

Examples:

Example 1. Subtract. $3\frac{4}{5} - 1\frac{3}{4} =$

Solution: Convert mixed numbers into fractions: $3\frac{4}{5} = \frac{3\times5+4}{5} = \frac{19}{5}$ and $1\frac{3}{4} = \frac{1\times4+3}{4} = \frac{7}{4}$

These two fractions are "unlike" fractions. (they have different denominators). Find equivalent fractions with the same denominator. Use this formula: $\frac{a}{b} - \frac{c}{d} = \frac{ad-bc}{bd}$

$\frac{19}{5} - \frac{7}{4} = \frac{(19)(4)-(5)(7)}{5\times4} = \frac{76-35}{20} = \frac{41}{20}$, the answer is an improper fraction, convert it into a mixed number. $\frac{41}{20} = 2\frac{1}{20}$

Example 2. Subtract. $4\frac{3}{8} - 1\frac{1}{2} =$

Solution: Convert mixed numbers into fractions: $4\frac{3}{8} = \frac{4\times8+3}{8} = \frac{35}{8}$ and $1\frac{1}{2} = \frac{1\times2+1}{4} = \frac{3}{2}$

Find equivalent fractions: $\frac{3}{2} = \frac{12}{8}$. Then: $4\frac{3}{8} - 1\frac{1}{2} = \frac{35}{8} - \frac{12}{8} = \frac{23}{8}$

The answer is an improper fraction, convert it into a mixed number.
$$\frac{23}{8} = 2\frac{7}{8}$$

MULTIPLYING MIXED NUMBERS

Use the following steps for multiplying mixed numbers:

☑ Convert the mixed numbers into fractions. $a\frac{c}{b} = a + \frac{c}{b} = \frac{ab+c}{b}$

☑ Multiply fractions. $\frac{a}{b} \times \frac{c}{d} = \frac{a \times c}{b \times d}$

☑ Write your answer in lowest terms.

☑ If the answer is an improper fraction (numerator is bigger than denominator), convert it into a mixed number.

Examples:

Example 1. Multiply. $3\frac{1}{3} \times 4\frac{1}{6} =$

Solution: Convert mixed numbers into fractions, $3\frac{1}{3} = \frac{3 \times 3 + 1}{3} = \frac{10}{3}$ and $4\frac{1}{6} = \frac{4 \times 6 + 1}{6} = \frac{25}{6}$

Apply the fractions rule for multiplication, $\frac{10}{3} \times \frac{25}{6} = \frac{10 \times 25}{3 \times 6} = \frac{250}{18}$

The answer is an improper fraction. Convert it into a mixed number. $\frac{250}{18} = 13\frac{8}{9}$

Example 2. Multiply. $2\frac{1}{2} \times 3\frac{2}{3} =$

Solution: Converting mixed numbers into fractions, $2\frac{1}{2} \times 3\frac{2}{3} = \frac{5}{2} \times \frac{11}{3}$

Apply the fractions rule for multiplication, $\frac{5}{2} \times \frac{11}{3} = \frac{5 \times 11}{2 \times 3} = \frac{55}{6} = 9\frac{1}{6}$

Example 3. Multiply mixed numbers. $2\frac{1}{3} \times 2\frac{1}{2} =$

Solution: Converting mixed numbers to fractions, $2\frac{1}{3} = \frac{7}{3}$ and $2\frac{1}{2} = \frac{5}{2}$. Multiply two fractions:

$$\frac{7}{3} \times \frac{5}{2} = \frac{7 \times 5}{3 \times 2} = \frac{35}{6} = 5\frac{5}{6}$$

DIVIDING MIXED NUMBERS

Use the following steps for dividing mixed numbers:

☑ Convert the mixed numbers into fractions. $a\frac{c}{b} = a + \frac{c}{b} = \frac{ab+c}{b}$

☑ Divide fractions: Keep, Change, Flip: Keep the first fraction, change the division sign to multiplication, and flip the numerator and denominator of the second fraction. Then, solve! $\frac{a}{b} \div \frac{c}{d} = \frac{a}{b} \times \frac{d}{c} = \frac{a \times d}{b \times c}$

☑ Write your answer in lowest terms.

☑ If the answer is an improper fraction (numerator is bigger than denominator), convert it into a mixed number.

Examples:

Example 1. Solve. $3\frac{2}{3} \div 2\frac{1}{2}$

Solution: Convert mixed numbers into fractions: $3\frac{2}{3} = \frac{3\times3+2}{3} = \frac{11}{3}$ and $2\frac{1}{2} = \frac{2\times2+1}{2} = \frac{5}{2}$

Keep, Change, Flip: $\frac{11}{3} \div \frac{5}{2} = \frac{11}{3} \times \frac{2}{5} = \frac{11\times2}{3\times5} = \frac{22}{15}$. The answer is an improper fraction. Convert it into a mixed number: $\frac{22}{15} = 1\frac{7}{15}$

Example 2. Solve. $3\frac{4}{5} \div 1\frac{5}{6}$

Solution: Convert mixed numbers to fractions, then solve:
$3\frac{4}{5} \div 1\frac{5}{6} = \frac{19}{5} \div \frac{11}{6} = \frac{19}{5} \times \frac{6}{11} = \frac{114}{55} = 2\frac{4}{55}$

Example 3. Solve. $2\frac{2}{7} \div 2\frac{3}{5}$

Solution: Converting mixed numbers to fractions: $3\frac{4}{5} \div 1\frac{5}{6} = \frac{16}{7} \div \frac{13}{5}$

Keep, Change, Flip: $\frac{16}{7} \div \frac{13}{5} = \frac{16}{7} \times \frac{5}{13} = \frac{16\times5}{7\times13} = \frac{80}{91}$

CHAPTER 1: PRACTICES

✍ Simplify each fraction.

1) $\dfrac{16}{24} =$

2) $\dfrac{28}{70} =$

3) $\dfrac{30}{105} =$

4) $\dfrac{40}{35} =$

5) $\dfrac{48}{56} =$

6) $\dfrac{6}{120} =$

7) $\dfrac{15}{100} =$

8) $\dfrac{45}{54} =$

✍ Find the sum or difference.

9) $\dfrac{4}{12} + \dfrac{3}{12} =$

10) $\dfrac{5}{4} + \dfrac{1}{12} =$

11) $\dfrac{3}{6} + \dfrac{2}{5} =$

12) $\dfrac{8}{25} - \dfrac{3}{25} =$

13) $\dfrac{5}{3} - \dfrac{2}{9} =$

14) $\dfrac{3}{2} - \dfrac{3}{4} =$

15) $\dfrac{4}{3} - \dfrac{6}{5} =$

16) $\dfrac{5}{12} + \dfrac{3}{5} =$

✍ Find the products or quotients.

17) $\dfrac{9}{5} \div \dfrac{3}{2} =$

18) $\dfrac{8}{7} \div \dfrac{4}{3} =$

19) $\dfrac{6}{4} \times \dfrac{8}{5} =$

20) $\dfrac{7}{2} \times \dfrac{4}{9} =$

✍ Find the sum.

21) $2\dfrac{1}{3} + 1\dfrac{4}{5} =$

22) $4\dfrac{3}{7} + 3\dfrac{3}{4} =$

23) $2\dfrac{3}{4} + 3\dfrac{1}{3} =$

24) $1\dfrac{1}{4} + 3\dfrac{1}{2} =$

25) $2\dfrac{5}{7} + 2\dfrac{1}{3} =$

26) $4\dfrac{2}{9} + 2\dfrac{1}{2} =$

✍ Find the difference.

27) $4\frac{2}{9} - 3\frac{1}{7} =$

28) $3\frac{3}{4} - 2\frac{1}{8} =$

29) $3\frac{2}{7} - 2\frac{4}{9} =$

30) $8\frac{3}{4} - 2\frac{1}{8} =$

31) $5\frac{5}{6} - 3\frac{1}{24} =$

32) $7\frac{3}{10} - 4\frac{4}{5} =$

33) $8\frac{1}{6} - 3\frac{2}{3} =$

34) $14\frac{9}{10} - 8\frac{4}{5} =$

✍ Find the products.

35) $2\frac{1}{9} \times 2\frac{5}{6} =$

36) $2\frac{3}{4} \times 4\frac{1}{9} =$

37) $1\frac{2}{7} \times 1\frac{5}{6} =$

38) $3\frac{2}{9} \times 1\frac{6}{5} =$

39) $3\frac{2}{3} \times 2\frac{3}{5} =$

40) $2\frac{5}{6} \times 3\frac{1}{9} =$

41) $3\frac{4}{5} \times 1\frac{1}{6} =$

42) $4\frac{1}{5} \times 1\frac{2}{7} =$

✍ Solve.

43) $8\frac{3}{4} \div 4\frac{1}{3} =$

44) $4\frac{2}{5} \div 1\frac{2}{9} =$

45) $6\frac{1}{2} \div 2\frac{1}{3} =$

46) $7\frac{1}{6} \div 3\frac{4}{9} =$

47) $2\frac{1}{4} \div 1\frac{1}{8} =$

48) $3\frac{2}{5} \div 1\frac{1}{10} =$

49) $4\frac{1}{2} \div 2\frac{2}{3} =$

50) $11\frac{1}{3} \div 2\frac{2}{9} =$

CHAPTER 1: ANSWERS

1) $\frac{2}{3}$

2) $\frac{2}{5}$

3) $\frac{2}{7}$

4) $\frac{8}{7}$

5) $\frac{6}{7}$

6) $\frac{1}{20}$

7) $\frac{3}{20}$

8) $\frac{5}{6}$

9) $\frac{7}{12}$

10) $\frac{4}{3}$

11) $\frac{9}{10}$

12) $\frac{1}{5}$

13) $\frac{13}{9} = 1\frac{4}{9}$

14) $\frac{3}{4}$

15) $\frac{2}{15}$

16) $\frac{61}{60} = 1\frac{1}{60}$

17) $\frac{6}{5}$

18) $\frac{6}{7}$

19) $\frac{12}{5} = 2\frac{2}{5}$

20) $\frac{14}{9} = 1\frac{5}{9}$

21) $4\frac{2}{15}$

22) $8\frac{5}{28}$

23) $6\frac{1}{12}$

24) $4\frac{3}{4}$

25) $5\frac{1}{21}$

26) $6\frac{13}{18}$

27) $1\frac{5}{63}$

28) $1\frac{5}{8}$

29) $\frac{53}{63}$

30) $6\frac{5}{8}$

31) $2\frac{19}{24}$

32) $2\frac{1}{2}$

33) $4\frac{1}{2}$

34) $6\frac{1}{10}$

35) $5\frac{53}{54}$

36) $11\frac{11}{36}$

37) $2\frac{5}{14}$

38) $7\frac{4}{45}$

39) $9\frac{8}{15}$

40) $8\frac{22}{27}$

41) $4\frac{13}{30}$

42) $5\frac{2}{5}$

43) $2\frac{1}{52}$

44) $3\frac{3}{5}$

45) $2\frac{11}{14}$

46) $2\frac{5}{62}$

47) 2

48) $3\frac{1}{11}$

49) $1\frac{11}{16}$

50) $5\frac{1}{10}$

CHAPTER 2:
DECIMALS

Math Topics that you'll learn in this chapter:

▶ Comparing Decimals

▶ Rounding Decimals

▶ Adding and Subtracting Decimals

▶ Multiplying and Dividing Decimals

COMPARING DECIMALS

☑ A decimal is a fraction written in a special form. For example, instead of writing $\frac{1}{2}$ you can write 0.5

☑ A Decimal Number contains a Decimal Point. It separates the whole number part from the fractional part of a decimal number.

☑ Let's review decimal place values: Example: **53.9861**

5: tens   3: ones   9: tenths

8: hundredths   6: thousandths   1: tens thousandths

☑ To compare decimals, compare each digit of two decimals in the same place value. Start from left. Compare hundreds, tens, ones, tenth, hundredth, etc.

☑ To compare numbers, use these symbols:

Equal to =,   Less than <,   Greater than >

Greater than or equal ≥,   Less than or equal ≤

Examples:

Example 1. Compare 0.60 and 0.06.

Solution: 0.60 *is greater than* 0.06, because the tenth place of 0.60 is 6, but the tenth place of 0.06 is zero. Then: 0.60 > 0.06

Example 2. Compare 0.0815 and 0.815.

Solution: 0.815 *is greater than* 0.0815, because the tenth place of 0.815 is 8, but the tenth place of 0.0815 is zero. Then: 0.0815 < 0.815

ROUNDING DECIMALS

☑ We can round decimals to a certain accuracy or number of decimal places. This is used to make calculations easier to do and results easier to understand when exact values are not too important.

☑ First, you'll need to remember your place values: For example: **12.4869**

| 1: tens | 2: ones | 4: tenths |
| 8: hundredths | 6: thousandths | 9: tens thousandths |

☑ To round a decimal, first find the place value you'll round to.

☑ Find the digit to the right of the place value you're rounding to. If it is 5 or bigger, add 1 to the place value you're rounding to and remove all digits on its right side. If the digit to the right of the place value is less than 5, keep the place value and remove all digits on the right.

Examples:

Example 1. Round 1.9278 to the thousandth place value.

Solution: First, look at the next place value to the right, (tens thousandths). It's 8 and it is greater than 5. Thus add 1 to the digit in the thousandth place. The thousandth place is 7. → 7 + 1 = 8, then,
The answer is 1.928

Example 2. Round 9.4126 to the nearest hundredth.

Solution: First, look at the digit to the right of hundredth (thousandths place value). It's 2 and it is less than 5, thus remove all the digits to the right of hundredth place. Then, the answer is 9.41

ADDING AND SUBTRACTING DECIMALS

☑ Line up the decimal numbers.

☑ Add zeros to have the same number of digits for both numbers if necessary.

☑ Remember your place values: For example: 73.5196

    7: tens   3: ones   5: tenths

    1: hundredths   9: thousandths   6: tens thousandths

☑ Add or subtract using column addition or subtraction.

Examples:

Example 1. Add. $1.8 + 3.12$

Solution: First, line up the numbers: $\begin{array}{r} 1.8 \\ +\,3.12 \\ \hline \end{array}$ → Add a zero to have the same number of digits for both numbers. $\begin{array}{r} 1.80 \\ +\,3.12 \\ \hline \end{array}$ → Start with the hundredths place: $0 + 2 = 2$, $\begin{array}{r} 1.80 \\ +\,3.12 \\ \hline 2 \end{array}$ → Continue with tenths place: $8 + 1 = 9$, $\begin{array}{r} 1.80 \\ +\,3.12 \\ \hline .92 \end{array}$ → Add the ones place: $3 + 1 = 4$, $\begin{array}{r} 1.80 \\ +\,3.12 \\ \hline 4.92 \end{array}$

Example 2. Find the difference. $3.67 - 2.23$

Solution: First, line up the numbers: $\begin{array}{r} 3.67 \\ -\,2.23 \\ \hline \end{array}$ → Start with the hundredths place: $7 - 3 = 4$, $\begin{array}{r} 3.67 \\ -\,2.23 \\ \hline 4 \end{array}$ → Continue with tenths place. $6 - 2 = 4$, $\begin{array}{r} 3.67 \\ -\,2.23 \\ \hline .44 \end{array}$ → Subtract the ones place. $3 - 2 = 1$, $\begin{array}{r} 3.67 \\ -\,2.23 \\ \hline 1.44 \end{array}$

MULTIPLYING AND DIVIDING DECIMALS

For multiplying decimals:

☑ Ignore the decimal point and set up and multiply the numbers as you do with whole numbers.

☑ Count the total number of decimal places in both of the factors.

☑ Place the decimal point in the product.

For dividing decimals:

☑ If the divisor is not a whole number, move the decimal point to the right to make it a whole number. Do the same for the dividend.

☑ Divide similar to whole numbers.

Examples:

Example 1. Find the product. $0.81 \times 0.32 =$

Solution: Set up and multiply the numbers as you do with whole numbers. Line up the numbers: $\begin{array}{r} 81 \\ \times 32 \\ \hline \end{array}$ → Start with the ones place then continue with other digits → $\begin{array}{r} 81 \\ \times 32 \\ \hline 2,592 \end{array}$. Count the total number of decimal places in both of the factors. There are four decimals digits. (two for each factor 0.81 and 0.32) Then: $0.81 \times 0.32 = 0.2592$

Example 2. Find the quotient. $1.60 \div 0.4 =$

Solution: The divisor is not a whole number. Multiply it by 10 to get 4: → $0.4 \times 10 = 4$
Do the same for the dividend to get 16. → $1.60 \times 10 = 1.6$
Now, divide $16 \div 4 = 4$. The answer is 4.

CHAPTER 2: PRACTICES

✎ Compare. Use >, =, and <

1) 0.55 ☐ 0.055

2) 0.34 ☐ 0.33

3) 0.66 ☐ 0.59

4) 2.650 ☐ 2.65

5) 2.34 ☐ 2.67

6) 2.46 ☐ 2.05

7) 0.16 ☐ 0.025

8) 5.05 ☐ 50.5

9) 1.020 ☐ 1.02

10) 3.022 ☐ 3.3

11) 1.400 ☐ 1.60

12) 3.44 ☐ 4.3

13) 0.380 ☐ 3.03

14) 2.081 ☐ 2.63

✎ Round each decimal to the nearest whole number.

15) 10.57

16) 4.8

17) 29.7

18) 32.58

19) 7.5

20) 8.87

21) 56.23

22) 6.39

23) 18.63

24) 25.56

25) 28.49

26) 12.67

27) 49.9

28) 17.77

29) 3.44

30) 55.56

✑ Find the sum or difference.

31) $25.31 + 56.37 =$

32) $78.32 - 65.10 =$

33) $65.80 + 14.26 =$

34) $90.24 - 53.81 =$

35) $76.41 - 49.27 =$

36) $45.39 + 17.86 =$

37) $56.02 + 30.60 =$

38) $67.01 - 28.40 =$

39) $75.14 - 25.96 =$

40) $37.52 + 13.50 =$

41) $84.71 - 54.18 =$

42) $24.12 + 29.84 =$

43) $50.59 - 46.25 =$

44) $63.13 + 21.14 =$

45) $45.23 - 35.17 =$

46) $18.02 + 30.40 =$

✑ Find the product or quotient.

47) $1.4 \times 3.2 =$

48) $8.2 \div 0.2 =$

49) 4.12×3.5

50) $6.8 \div 1.7 =$

51) $5.8 \times 0.5 =$

52) $1.54 \div 0.5 =$

53) $1.4 \times 3.2 =$

54) $5.8 \div 0.2 =$

55) $6.4 \times 7.3 =$

56) $0.3 \times 3.2 =$

57) $7.5 \times 5.6 =$

58) $45.6 \div 0.8 =$

59) $1.9 \times 5.8 =$

60) $6.74 \times 2.5 =$

61) $56.08 \div 0.2 =$

62) $36.2 \times 3.6 =$

CHAPTER 2: ANSWERS

1) >

2) >

3) >

4) =

5) <

6) >

7) >

8) <

9) =

10) <

11) <

12) <

13) <

14) <

15) 11

16) 5

17) 30

18) 33

19) 8

20) 9

21) 56

22) 6

23) 19

24) 26

25) 28

26) 13

27) 50

28) 18

29) 3

30) 56

31) 81.68

32) 13.22

33) 80.06

34) 36.43

35) 27.14

36) 63.25

37) 86.62

38) 38.61

39) 49.18

40) 51.02

41) 30.53

42) 53.96

43) 4.34

44) 84.27

45) 10.06

46) 48.42

47) 4.48

48) 41

49) 14.42

50) 4

51) 2.9

52) 3.08

53) 4.48

54) 29

55) 46.72

56) 0.96

57) 42

58) 57

59) 11.02

60) 16.85

61) 280.4

62) 130.32

CHAPTER 3:

INTEGERS AND ORDER OF OPERATIONS

Math Topics that you'll learn in this chapter:

▶ Adding and Subtracting Integers

▶ Multiplying and Dividing Integers

▶ Order of Operations

▶ Integers and Absolute Value

ADDING AND SUBTRACTING INTEGERS

- ☑ Integers include zero, counting numbers, and the negative of the counting numbers. $\{\ldots, -3, -2, -1, 0, 1, 2, 3, \ldots\}$

- ☑ Add a positive integer by moving to the right on the number line. (you will get a bigger number)

- ☑ Add a negative integer by moving to the left on the number line. (you will get a smaller number)

- ☑ Subtract an integer by adding its opposite.

Examples:

Example 1. Solve. $(-4) - 5 =$

Solution: Keep the first number and convert the sign of the second number to its opposite. (change subtraction into addition. Then: $(-4) + 5 = 1$

Example 2. Solve. $11 + (8 - 19) =$

Solution: First, subtract the numbers in brackets, $8 - 19 = -11$.
Then: $11 + (-11) = \rightarrow$ change addition into subtraction: $11 - 11 = 0$

Example 3. Solve. $5 - 14 - 3 =$

Solution: First, subtract the numbers in brackets, $-14 - 3 = -17$
Then: $5 - 17 = \rightarrow$ change subtraction into addition: $5 + 17 = 22$

Example 4. Solve. $10 + (-6 - 15) =$

Solution: First, subtract the numbers in brackets, $-6 - 15 = -21$
Then: $10 + (-21) = \rightarrow$ change addition into subtraction: $10 - 21 = -11$

MULTIPLYING AND DIVIDING INTEGERS

Use the following rules for multiplying and dividing integers:

☑ (negative) × (negative) = positive

☑ (negative) ÷ (negative) = positive

☑ (negative) × (positive) = negative

☑ (negative) ÷ (positive) = negative

☑ (positive) × (positive) = positive

☑ (positive) ÷ (negative) = negative

Examples:

Example 1. Solve. $2 \times (-3) =$

Solution: Use this rule: (positive) × (negative) = negative.
Then: $(2) \times (-3) = -6$

Example 2. Solve. $(-5) + (-27 \div 9) =$

Solution: First, divide -27 by 9, the numbers in brackets, use this rule:
(negative) ÷ (positive) = negative. Then: $-27 \div 9 = -3$
$(-5) + (-27 \div 9) = (-5) + (-3) = -5 - 3 = -8$

Example 3. Solve. $(15 - 17) \times (-8) =$

Solution: First, subtract the numbers in brackets,
$15 - 17 = -2 \rightarrow (-2) \times (-8) =$
Now use this rule: (negative) × (negative) = positive $\rightarrow (-2) \times (-8) = 16$

Example 4. Solve. $(16 - 10) \div (-2) =$

Solution: First, subtract the numbers in brackets,
$16 - 10 = 6 \rightarrow (6) \div (-2) =$
Now use this rule: (positive) ÷ (negative) = negative $\rightarrow (6) \div (-2) = -3$

ORDER OF OPERATIONS

☑ In Mathematics, "operations" are addition, subtraction, multiplication, division, exponentiation (written as b^n), and grouping;

☑ When there is more than one math operation in an expression, use PEMDAS: (to memorize this rule, remember the phrase "Please Excuse My Dear Aunt Sally".)

❖ Parentheses

❖ Exponents

❖ Multiplication and Division (from left to right)

❖ Addition and Subtraction (from left to right)

Examples:

Example 1. Calculate. $(3 + 5) \div (3^2 \div 9) =$

Solution: First, simplify inside parentheses:
$(8) \div (9 \div 9) = (8) \div (1)$, Then: $(8) \div (1) = 8$

Example 2. Solve. $(7 \times 8) - (12 - 4) =$

Solution: First, calculate within parentheses: $(7 \times 8) - (12 - 4) = (56) - (8)$, Then: $(56) - (8) = 48$

Example 3. Calculate. $-2[(8 \times 9) \div (2^2 \times 2)] =$

Solution: First, calculate within parentheses:
$-2[(72) \div (4 \times 2)] = -2[(72) \div (8)] = -2[9]$
multiply -2 and 9. Then: $-2[9] = -18$

Example 4. Solve. $(14 \div 7) + (-13 + 8) =$

Solution: First, calculate within parentheses:
$(14 \div 7) + (-13 + 8) = (2) + (-5)$ Then: $(2) - (5) = -3$

INTEGERS AND ABSOLUTE VALUE

☑ The absolute value of a number is its distance from zero, in either direction, on the number line. For example, the distance of 9 and −9 from zero on number line is 9.

☑ The absolute value of an integer is the numerical value without its sign. (negative or positive)

☑ The vertical bar is used for absolute value as in $|x|$.

☑ The absolute value of a number is never negative; because it only shows, "how far the number is from zero".

Examples:

Example 1. Calculate. $|12 − 4| × 4 =$

Solution: First, solve $|12 − 4|$, $→|12 − 4| = |8|$, the absolute value of 8 is 8, $|8| = 8$ Then: $8 × 4 = 32$

Example 2. Solve. $\frac{|−16|}{4} × |3 − 8| =$

Solution: First, find $|−16|$, $→$ the absolute value of −16 is 16,
Then: $|−16| = 16$, $\frac{16}{4} × |3 − 8| =$

Now, calculate $|3 − 8|$, $→$ $|3 − 8| = |−5|$, the absolute value of −5 is 5. $|−5| = 5$ then: $\frac{16}{4} × 5 = 4 × 5 = 20$

Example 3. Solve. $|9 − 3| × \frac{|−3×8|}{6} =$

Solution: First, calculate $|9 − 3|$,$→|9 − 3| = |6|$, the absolute value of 6 is 6, $|6| = 6$. Then: $6 × \frac{|−3×8|}{6}$

Now calculate $|−3 × 8|$, $→$ $|−3 × 8| = |−24|$, the absolute value of −24 is 24, $|−24| = 24$ Then: $6 × \frac{24}{6} = 6 × 4 = 24$

CHAPTER 3: PRACTICES

✍ **Find each sum or difference.**

1) $13 - (-6) =$

2) $(-36) + 18 =$

3) $(-6) + (-22) =$

4) $54 + (-12) + 9 =$

5) $35 + (-24 + 4) =$

6) $(-17) + (-34 + 12) =$

7) $(-1) + (28 - 15) =$

8) $5 + (-9 + 12) =$

9) $(-10) + (-20) =$

10) $32 - (-5) =$

11) $(-9) + (24 - 3) =$

12) $8 - (-2 + 12) =$

13) $(-3) + (45 + 3) =$

14) $5 + (-30 + 6) =$

15) $(-6 + 1) + (-20) =$

16) $(-7) - (-20 + 2) =$

17) $(-6) - (2) =$

18) $(9 - 6) - (-3) =$

✍ **Solve.**

19) $4 \times (-8) =$

20) $(-27) \div (-9) =$

21) $(-2) \times (-9) \times 3 =$

22) $5 \times (-3) \times (-7) =$

23) $(-10 - 8) \div (-9) =$

24) $(-9 + 7) \times (-20) =$

25) $(-7) \times (-5) =$

26) $(-6) \times (-2 + 6) =$

27) $(-3) \times (-4) \times 3 =$

28) $(-8 - 2) \times (-1 + 4) =$

29) $(-9) \times (-20) =$

30) $6 \times (-2 + 9) =$

31) $(-5 - 6) \times (-2) =$

32) $(-4 - 2) \times (-3 - 7) =$

33) $(-9) \div (13 - 16) =$

34) $56 \times (-8) =$

35) $(-9 - 3) \div (-4) =$

36) $72 \div (-18 + 10) =$

✍ Evaluate each expression.

37) $2 + (6 \times 4) =$

38) $(7 \times 9) - 8 =$

39) $(-6) + (2 \times 9) =$

40) $(-2 - 4) + (3 \times 7) =$

41) $(28 \div 7) - (5 \times 3) =$

42) $(9 \times 3) + (6 \times 4) =$

43) $(36 \div 4) - (36 \div 6) =$

44) $(7 + 3) + (16 \div 2) =$

45) $(15 \times 3) - 16 =$

46) $8 - (7 \times 3) =$

47) $(9 + 15) \div (8 \div 4) =$

48) $2[(3 \times 6) + (16 \times 2)] =$

49) $(18 - 6) + (4 \times 2) =$

50) $2[(2 \times 3) - (8 \times 5)] =$

51) $(9 + 7) \div (16 \div 8) =$

52) $(3 + 9) \times (25 \times 2) =$

53) $3[(10 \times 9) \div (9 \times 5)] =$

54) $-6[(10 \times 9) \div (5 \times 6)] =$

✍ Find the answers.

55) $|-8| + |6 - 15| =$

56) $|-5 + 9| + |-3| =$

57) $|-6| + |2 - 10| =$

58) $|-8 + 3| - |4 - 8| =$

59) $|6 - 10| + |5 - 7| =$

60) $|-6| - |-9 - 19| + 5 =$

61) $|-9 + 2| - |3 - 5| + 6 =$

62) $4 + |3 - 7| + |2 - 6| =$

63) $\frac{|-64|}{8} \times \frac{|-48|}{6} =$

64) $\frac{|-36|}{6} \times \frac{|-56|}{8} =$

65) $\frac{|-72|}{9} \times \frac{|-45|}{5} =$

66) $|8 \times (-1)| \times \frac{|-24|}{3} =$

67) $|-2 \times 5| \times \frac{|-27|}{9} =$

68) $\frac{|-144|}{12} - |-6 \times 4| =$

69) $\frac{|-63|}{7} + |-9 \times 2| =$

70) $\frac{|-70|}{7} + |-8 \times 3| =$

71) $\frac{|-7 \times -3|}{7} \times \frac{|8 \times (-5)|}{8} =$

72) $\frac{|(-2) \times (-6)|}{4} \times \frac{|8 \times (-4)|}{2} =$

CHAPTER 3: ANSWERS

1) 19	25) 35	49) 20
2) −18	26) −24	50) −68
3) −28	27) 36	51) 8
4) 51	28) −30	52) 600
5) 15	29) 180	53) 6
6) −39	30) 42	54) −18
7) 12	31) 22	55) 17
8) 8	32) 80	56) 7
9) −30	33) 3	57) 14
10) 37	34) −448	58) 1
11) 12	35) 3	59) 6
12) −2	36) −9	60) −17
13) 45	37) 26	61) 11
14) −19	38) 55	62) 12
15) −25	39) 12	63) 64
16) 11	40) 15	64) 42
17) −8	41) −11	65) 72
18) 6	42) 51	66) 64
19) −32	43) 3	67) 30
20) 3	44) 18	68) −12
21) 54	45) 29	69) 27
22) 105	46) −13	70) 34
23) 2	47) 12	71) 15
24) 40	48) 100	72) 48

CHAPTER 4:

RATIOS AND PROPORTIONS

Math Topics that you'll learn in this chapter:

► Simplifying Ratios
► Proportional Ratios
► Similarity and Ratios

SIMPLIFYING RATIOS

☑ Ratios are used to make comparisons between two numbers.

☑ Ratios can be written as a fraction, using the word "to", or with a colon. Example: $\frac{3}{4}$ or "3 to 4" or 3:4

☑ You can calculate equivalent ratios by multiplying or dividing both sides of the ratio by the same number.

Examples:

Example 1. Simplify. $9:3 =$

Solution: Both numbers 9 and 3 are divisible by 3 , $\Rightarrow 9 \div 3 = 3$, $3 \div 3 = 1$, Then: $9:3 = 3:1$

Example 2. Simplify. $\frac{24}{44} =$

Solution: Both numbers 24 and 44 are divisible by 4, $\Rightarrow$ $24 \div 4 = 6$, $44 \div 4 = 11$, Then: $\frac{24}{44} = \frac{6}{11}$

Example 3. There are 36 students in a class and 16 are girls. Write the ratio of girls to boys.

Solution: Subtract 16 from 36 to find the number of boys in the class. $36 - 16 = 20$. There are 20 boys in the class. So, the ratio of girls to boys is $16:20$. Now, simplify this ratio. Both 20 and 16 are divisible by 4. Then: $20 \div 4 = 5$, and $16 \div 4 = 4$. In the simplest form, this ratio is $4:5$

Example 4. A recipe calls for butter and sugar in the ratio $3:4$. If you're using 9 cups of butter, how many cups of sugar should you use?

Solution: Since you use 9 cups of butter, or 3 times as much, you need to multiply the amount of sugar by 3. Then: $4 \times 3 = 12$. So, you need to use 12 cups of sugar. You can solve this using equivalent fractions: $\frac{3}{4} = \frac{9}{12}$

PROPORTIONAL RATIOS

☑ Two ratios are proportional if they represent the same relationship.

☑ A proportion means that two ratios are equal. It can be written in two ways:
$$\frac{a}{b} = \frac{c}{d} \qquad a : b = c : d$$

☑ The proportion $\frac{a}{b} = \frac{c}{d}$ can be written as: $a \times d = c \times b$

Examples:

Example 1. Solve this proportion for x. $\quad \frac{3}{7} = \frac{12}{x}$

Solution: Use cross multiplication: $\frac{3}{7} = \frac{12}{x} \Rightarrow 3 \times x = 7 \times 12 \Rightarrow 3x = 84$

Divide both sides by 3 to find x: $\qquad x = \frac{84}{3} \Rightarrow x = 28$

Example 2. If a box contains red and blue balls in ratio of $3 : 7$ red to blue, how many red balls are there if 49 blue balls are in the box?

Solution: Write a proportion and solve. $\frac{3}{7} = \frac{x}{49}$

Use cross multiplication: $\quad 3 \times 49 = 7 \times x \Rightarrow 147 = 7x$

Divide to find x: $x = \frac{147}{7} \Rightarrow x = 21$. There are 21 red balls in the box.

Example 3. Solve this proportion for x. $\quad \frac{2}{9} = \frac{12}{x}$

Solution: Use cross multiplication: $\frac{2}{9} = \frac{12}{x} \Rightarrow 2 \times x = 9 \times 12 \Rightarrow 2x = 108$

Divide to find x: $x = \frac{108}{2} \Rightarrow x = 54$

Example 4. Solve this proportion for x. $\frac{6}{7} = \frac{18}{x}$

Solution: Use cross multiplication: $\frac{6}{7} = \frac{18}{x} \Rightarrow 6 \times x = 7 \times 18 \Rightarrow 6x = 126$

Divide to find x: $x = \frac{126}{6} \Rightarrow x = 21$

SIMILARITY AND RATIOS

☑ Two figures are similar if they have the same shape.

☑ Two or more figures are similar if the corresponding angles are equal, and the corresponding sides are in proportion.

Examples:

Example 1. The following triangles are similar. What is the value of the unknown side?

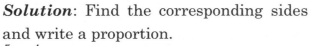

Solution: Find the corresponding sides and write a proportion.
$\frac{5}{10} = \frac{4}{x}$. Now, use the cross product to solve for x:
$\frac{5}{10} = \frac{4}{x} \rightarrow 5 \times x = 10 \times 4 \rightarrow 5x = 40$. Divide both sides by 5. Then: $5x = 40 \rightarrow \frac{5x}{5} = \frac{40}{5} \rightarrow x = 8$
The missing side is 8.

Example 2. Two rectangles are similar. The first is 6 feet wide and 20 feet long. The second is 15 feet wide. What is the length of the second rectangle?

Solution: Let's put x for the length of the second rectangle. Since two rectangles are similar, their corresponding sides are in proportion. Write a proportion and solve for the missing number.
$\frac{6}{15} = \frac{20}{x} \rightarrow 6x = 15 \times 20 \rightarrow 6x = 300 \rightarrow x = \frac{300}{6} = 50$
The length of the second rectangle is 50 feet.

CHAPTER 4: PRACTICES

✎ Reduce each ratio.

1) $3:21 = $ ___ : ___

2) $8:72 = $ ___ : ___

3) $21:49 = $ ___ : ___

4) $32:28 = $ ___ : ___

5) $35:45 = $ ___ : ___

6) $72:81 = $ ___ : ___

7) $36:54 = $ ___ : ___

8) $56:64 = $ ___ : ___

9) $12:36 = $ ___ : ___

10) $4:32 = $ ___ : ___

11) $16:48 = $ ___ : ___

12) $15:105 = $ ___ : ___

✎ Solve.

13) Bob has 18 red cards and 27 green cards. What is the ratio of Bob's red cards to his green cards? _____

14) In a party, 30 soft drinks are required for every 18 guests. If there are 240 guests, how many soft drinks are required? _____

15) Sara has 72 blue pens and 36 black pens. What is the ratio of Sara's black pens to her blue pens? _____

16) In Jack's class, 45 of the students are tall and 18 are short. In Michael's class 27 students are tall and 12 students are short. Which class has a higher ratio of tall to short students? _____

17) The price of 3 apples at the Quick Market is $1.44. The price of 5 of the same apples at Walmart is $2.45. Which place is the better buy? _____

18) The bakers at a Bakery can make 160 bagels in 4 hours. How many bagels can they bake in 14 hours? What is that rate per hour? _____

19) You can buy 5 cans of green beans at a supermarket for $3.40. How much does it cost to buy 35 cans of green beans? _____

✍ Solve each proportion.

20) $\frac{3}{4} = \frac{15}{x}, x =$

21) $\frac{9}{6} = \frac{x}{4}, x =$ _____

22) $\frac{3}{15} = \frac{2}{x}, x =$ _____

23) $\frac{5}{15} = \frac{3}{x}, x =$ _____

24) $\frac{24}{3} = \frac{x}{2}, x =$ _____

25) $\frac{8}{12} = \frac{10}{x}, x =$ _____

26) $\frac{3}{x} = \frac{2}{14}, x =$ _____

27) $\frac{10}{x} = \frac{3}{6}, x =$ _____

28) $\frac{15}{6} = \frac{x}{4}, x =$ _____

29) $\frac{x}{12} = \frac{5}{10}, x =$ _____

30) $\frac{18}{6} = \frac{3}{x}, x =$ _____

31) $\frac{3}{4} = \frac{24}{x}, x =$ _____

32) $\frac{8}{4} = \frac{x}{2}, x =$ _____

33) $\frac{12}{3} = \frac{x}{4}, x =$ _____

34) $\frac{24}{8} = \frac{x}{2}, x =$ _____

35) $\frac{5}{3} = \frac{x}{6}, x =$ _____

36) $\frac{10}{8} = \frac{x}{4}, x =$ _____

37) $\frac{x}{6} = \frac{6}{4}, x =$ _____

38) $\frac{x}{4} = \frac{7}{2}, x =$ _____

39) $\frac{9}{x} = \frac{3}{4}, x =$ _____

40) $\frac{10}{x} = \frac{1}{5}, x =$ _____

41) $\frac{9}{2} = \frac{x}{8}, x =$ _____

✍ Solve each problem.

42) Two rectangles are similar. The first is 6 *feet* wide and 24 *feet* long. The second is 10 *feet* wide. What is the length of the second rectangle? _____

43) Two rectangles are similar. One is 4.8 *meters* by 6 *meters*. The longer side of the second rectangle is 27 *meters*. What is the other side of the second rectangle? _____

CHAPTER 4: ANSWERS

1) $1:7$
2) $1:9$
3) $3:7$
4) $8:7$
5) $7:9$
6) $8:9$
7) $2:3$
8) $7:8$
9) $1:3$
10) $1:8$
11) $1:3$
12) $1:7$
13) $2:3$
14) 144
15) $1:2$
16) $Jack's\ class = \frac{45}{18} = \frac{5}{2}$

    $Michael's\ class = \frac{27}{12} = \frac{9}{4}$

Jack's class has a higher ratio of tall to short student

17) Quick Market
18) 560
19) $\$23.80$
20) 20

21) 6
22) 10
23) 9
24) 16
25) 15
26) 21
27) 20
28) 10
29) 6
30) 1
31) 32
32) 4
33) 16
34) 6
35) 10
36) 5
37) 9
38) 14
39) 12
40) 50
41) 36
42) 40
43) $21.6\ meters$

CHAPTER 5:

PERCENTAGE

Math Topics that you'll learn in this chapter:

► Percentage Calculations

► Percent Problems

► Percent of Increase and Decrease

► Discount, Tax and Tip

► Simple Interest

PERCENT PROBLEMS

☑ Percent is a ratio of a number and 100. It always has the same denominator, 100. The percent symbol is "%".

☑ Percent means "per 100". So, 20% is 20/100.

☑ In each percent problem, we are looking for the base, or part or the percent.

☑ Use these equations to find each missing section in a percent problem:

❖ Base = Part ÷ Percent

❖ Part = Percent × Base

❖ Percent = Part ÷ Base

Examples:

Example 1. What is 25% of 60?

Solution: In this problem, we have percent (25%) and base (60) and we are looking for the "part". Use this formula: *part = percent × base*.
Then: $part = 25\% \times 60 = \frac{25}{100} \times 60 = 0.25 \times 60 = 15$. The answer: 25% of 60 is 15.

Example 2. 20 is what percent of 400?

Solution: In this problem, we are looking for the percent. Use this equation: $Percent = Part \div Base \rightarrow Percent = 20 \div 400 = 0.05 = 5\%$.
Then: 20 is 5 percent of 400.

PERCENT OF INCREASE AND DECREASE

☑ Percent of change (increase or decrease) is a mathematical concept that represents the degree of change over time.

☑ To find the percentage of increase or decrease:

1. New Number – Original Number

2. The result ÷ Original Number × 100

☑ Or use this formula: Percent of change $= \frac{new\ number - original\ number}{original\ number} \times 100$

☑ Note: If your answer is a negative number, then this is a percentage decrease. If it is positive, then this is a percentage increase.

Examples:

Example 1. The price of a shirt increases from \$20 to \$30. What is the percentage increase?

Solution: First, find the difference: $30 - 20 = 10$

Then: $10 \div 20 \times 100 = \frac{10}{20} \times 100 = 50$. The percentage increase is 50. It means that the price of the shirt increased by 50%.

Example 2. The price of a table increased from \$25 to \$40. What is the percent of increase?

Solution: Use percentage formula:

$Percent\ of\ change = \dfrac{new\ number - original\ number}{original\ number} \times 100 =$

$\frac{40-25}{25} \times 100 = \frac{15}{25} \times 100 = 0.6 \times 100 = 60$. The percentage increase is 60. It means that the price of the table increased by 60%.

DISCOUNT, TAX AND TIP

☑ To find the discount: Multiply the regular price by the rate of discount

☑ To find the selling price: Original price − discount

☑ To find tax: Multiply the tax rate to the taxable amount (income, property value, etc.)

☑ To find the tip, multiply the rate to the selling price.

Examples:

Example 1. With an 10% discount, Ella saved $45 on a dress. What was the original price of the dress?

Solution: let x be the original price of the dress. Then: $10\% \ of \ x = 45$. Write an equation and solve for x: $0.10 \times x = 45 \rightarrow x = \frac{45}{0.10} = 450$. The original price of the dress was $450.

Example 2. Sophia purchased a new computer for a price of $950 at the Apple Store. What is the total amount her credit card is charged if the sales tax is 7%?

Solution: The taxable amount is $950, and the tax rate is 7%. Then: $Tax = 0.07 \times 950 = 66.50$
$Final \ price = Selling \ price + Tax \rightarrow final \ price = \$950 + \$66.50 = \$1,016.50$

Example 3. Nicole and her friends went out to eat at a restaurant. If their bill was $80.00 and they gave their server a 15% tip, how much did they pay altogether?

Solution: First, find the tip. To find the tip, multiply the rate to the bill amount. $Tip = 80 \times 0.15 = 12$. The final price is: $\$80 + \$12 = \$92$

SIMPLE INTEREST

☑ Simple Interest: The charge for borrowing money or the return for lending it.

☑ Simple interest is calculated on the initial amount (principal).

☑ To solve a simple interest problem, use this formula:

Interest = principal x rate x time $\quad(I = p \times r \times t = prt)$

Examples:

Example 1. Find simple interest for $300 investment at 6% for 5 years.

Solution: Use Interest formula:
$I = prt$ $(P = \$300,\ r = 6\% = \frac{6}{100} = 0.06$ and $t = 5)$
Then: $I = 300 \times 0.06 \times 5 = \$90$

Example 2. Find simple interest for $1,600 at 5% for 2 years.

Solution: Use Interest formula:
$I = prt$ $(P = \$1,600,\ r = 5\% = \frac{5}{100} = 0.05$ and $t = 2)$
Then: $I = 1,600 \times 0.05 \times 2 = \$160$

Example 3. Andy received a student loan to pay for his educational expenses this year. What is the interest on the loan if he borrowed $6,500 at 8% for 6 years?

Solution: Use Interest formula:$I = prt$. $P = \$6,500,\ r = 8\% = 0.08$ and $t = 6$
Then: $I = 6,500 \times 0.08 \times 8 = \$3,120$

Example 4. Bob is starting his own small business. He borrowed $10,000 from the bank at a 6% rate for 6 months. Find the interest Bob will pay on this loan.

Solution: Use Interest formula:
$I = prt$. $P = \$10,000,\ r = 6\% = 0.06$ and $t = 0.5$ (6 months is half year).
Then: $I = 10,000 \times 0.06 \times 0.5 = \$300$

CHAPTER 5: PRACTICES

✍ Solve each problem.

1) 10 is what percent of 80? ____%

2) 12 is what percent of 60? ____%

3) 20 is what percent of 80? ____%

4) 18 is what percent of 72? ____%

5) 16 is what percent of 50? ____%

6) 35 is what percent of 140? ____%

7) 12 is what percent of 240? ____%

8) 80 is what percent of 400? ____%

9) 60 is what percent of 300? ____%

10) 100 is what percent of 250? ____%

11) 25 is what percent of 400? ____%

12) 60 is what percent of 480? ____%

✍ Solve each problem.

13) Bob got a raise, and his hourly wage increased from $16 to $20. What is the percent increase? _____ %

14) The price of a pair of shoes increases from $30 to $36. What is the percent increase? ___ %

15) At a coffeeshop, the price of a cup of coffee increased from $1.30 to $1.56. What is the percent increase in the cost of the coffee? _____ %

16) A $40 shirt now selling for $28 is discounted by what percent? _____ %

17) Joe scored 20 out of 25 marks in Algebra, 30 out of 40 marks in science and 68 out of 80 marks in mathematics. In which subject his percentage of marks is best? _____

18) Emma purchased a computer for $408. The computer is regularly priced at $480. What was the percent discount Emma received on the computer? _____

19) A chemical solution contains 12% alcohol. If there is 42 ml of alcohol, what is the volume of the solution? _____

✒ Find the selling price of each item.

20) Original price of a computer: $700

   Tax: 9%, Selling price: $_____

21) Original price of a laptop: $460

   Tax: 20%, Selling price: $_____

22) Nicolas hired a moving company. The company charged $600 for its services, and Nicolas gives the movers a 12% tip. How much does Nicolas tip the movers? $_____

23) Mason has lunch at a restaurant and the cost of his meal is $50. Mason wants to leave a 25% tip. What is Mason's total bill, including tip? $_____

✒ Determine the simple interest for the following loans.

24) $480 *at* 6% *for* 5 *years.* $___

25) $500 *at* 5% *for* 3 *years.* $___

26) $360 *at* 3.5% *for* 2 *years.* $___

27) $600 at 4% for 4 years. $___

✒ Solve.

28) A new car, valued at $25,000, depreciates at 7% per year. What is the value of the car one year after purchase? $_____

29) Sara puts $6,000 into an investment yielding 4% annual simple interest; she left the money in for five years. How much interest does Sara get at the end of those five years? $_____

CHAPTER 5: ANSWERS

1) 12.5%

2) 20%

3) 25%

4) 25%

5) 32%

6) 25%

7) 5%

8) 20%

9) 20%

10) 40%

11) 6.25%

12) 12.5%

13) 25%

14) 20%

15) 20%

16) 30%

17) Mathematics

18) 15%

19) 350

20) $763

21) $552

22) $72

23) $62.50

24) $144

25) $75

26) $25.20

27) $96

28) $23,250

29) $1200

CHAPTER 6:

EXPRESSIONS AND VARIABLES

Math Topics that you'll learn in this chapter:

▶ Simplifying Variable Expressions

▶ Simplifying Polynomial Expressions

▶ The Distributive Property

▶ Evaluating One Variable

▶ Evaluating Two Variables

SIMPLIFYING VARIABLE EXPRESSIONS

☑ In algebra, a variable is a letter used to stand for a number. The most common letters are $x, y, z, a, b, c, m, and\ n$.

☑ An algebraic expression is an expression that contains integers, variables, and math operations such as addition, subtraction, multiplication, division, etc.

☑ In an expression, we can combine "like" terms. (values with same variable and same power)

Examples:

Example 1. Simplify. $(2x + 3x + 4) =$

Solution: In this expression, there are three terms: $2x, 3x$, and 4. Two terms are "like terms": $2x$ and $3x$. Combine like terms. $2x + 3x = 5x$. Then: $(2x + 3x + 4) = 5x + 4$ (**remember you cannot combine variables and numbers.**)

Example 2. Simplify. $12 - 3x^2 + 5x + 4x^2 =$

Solution: Combine "like" terms: $-3x^2 + 4x^2 = x^2$.
Then: $12 - 3x^2 + 5x + 4x^2 = 12 + x^2 + 5x$. Write in standard form (biggest powers first): $12 + x^2 + 5x = x^2 + 5x + 12$

Example 3. Simplify. $(10x^2 + 2x^2 + 3x) =$

Solution: Combine like terms. Then: $(10x^2 + 2x^2 + 3x) = 12x^2 + 3x$

Example 4. Simplify. $15x - 3x^2 + 9x + 5x^2 =$

Solution: Combine "like" terms: $15x + 9x = 24x$, and $-3x^2 + 5x^2 = 2x^2$
Then: $15x - 3x^2 + 9x + 5x^2 = 24x + 2x^2$. Write in standard form (biggest powers first): $24x + 2x^2 = 2x^2 + 24x$

SIMPLIFYING POLYNOMIAL EXPRESSIONS

☑ In mathematics, a polynomial is an expression consisting of variables and coefficients that involves only the operations of addition, subtraction, multiplication, and non–negative integer exponents of variables. $P(x) = a_n x^n + a_{n-1} x^{n-1} + \dots + a_2 x^2 + a_1 x + a_0$

☑ Polynomials must always be simplified as much as possible. It means you must add together any like terms. (values with same variable and same power)

Examples:

Example 1. Simplify this Polynomial Expressions. $x^2 - 5x^3 + 2x^4 - 4x^3$

Solution: Combine "like" terms: $-5x^3 - 4x^3 = -9x^3$
Then: $x^2 - 5x^3 + 2x^4 - 4x^3 = x^2 - 9x^3 + 2x^4$
Now, write the expression in standard form: $2x^4 - 9x^3 + x^2$

Example 2. Simplify this expression. $(2x^2 - x^3) - (x^3 - 4x^2) =$

Solution: First, use distributive property: $\rightarrow$ multiply $(-)$ into $(x^3 - 4x^2)$
$(2x^2 - x^3) - (x^3 - 4x^2) = 2x^2 - x^3 - x^3 + 4x^2$
Then combine "like" terms: $2x^2 - x^3 - x^3 + 4x^2 = 6x^2 - 2x^3$
And write in standard form: $6x^2 - 2x^3 = -2x^3 + 6x^2$

Example 3. Simplify. $4x^4 - 5x^3 + 15x^4 - 12x^3 =$

Solution: Combine "like" terms:
$-5x^3 - 12x^3 = -17x^3$ and $4x^4 + 15x^4 = 19x^4$
Then: $4x^4 - 5x^3 + 15x^4 - 12x^3 = 19x^4 - 17x^3$

THE DISTRIBUTIVE PROPERTY

☑ The distributive property (or the distributive property of multiplication over addition and subtraction) simplifies and solves expressions in the form of: $a(b + c)$ or $a(b - c)$

☑ The distributive property is multiplying a term outside the parentheses by the terms inside.

☑ Distributive Property rule: $a(b + c) = ab + ac$

Examples:

Example 1. Simply using the distributive property. $(-4)(x - 5)$

Solution: Use Distributive Property rule: $a(b + c) = ab + ac$
$(-4)(x - 5) = (-4 \times x) + (-4) \times (-5) = -4x + 20$

Example 2. Simply. $(3)(2x - 4)$

Solution: Use Distributive Property rule: $a(b + c) = ab + ac$
$(3)(2x - 4) = (3 \times 2x) + (3) \times (-4) = 6x - 12$

Example 3. Simply. $(-3)(3x - 5) + 4x$

Solution: First, simplify $(-3)(3x - 5)$ using the distributive property.
Then: $(-3)(3x - 5) = -9x + 15$
Now combine like terms: $(-3)(3x - 5) + 4x = -9x + 15 + 4x$
In this expression, $-9x$ and $4x$ are "like terms" and we can combine them.
$-9x + 4x = -5x$. Then: $-9x + 15 + 4x = -5x + 15$

EVALUATING ONE VARIABLE

☑ To evaluate one variable expression, find the variable and substitute a number for that variable.

☑ Perform the arithmetic operations.

Examples:

Example 1. Calculate this expression for x = 3. $15 - 3x$

Solution: First, substitute 3 for x
Then: $15 - 3x = 15 - 3(3)$
Now, use order of operation to find the answer: $15 - 3(3) = 15 - 9 = 6$

Example 2. Evaluate this expression for x = 1. $5x - 12$

Solution: First, substitute 1 for x,
Then: $5x - 12 = 5(1) - 12$
Now, use order of operation to find the answer: $5(1) - 12 = 5 - 12 = -7$

Example 3. Find the value of this expression when x = 5. $25 - 4x$

Solution: First, substitute 5 for x,
Then: $25 - 4x = 25 - 4(5) = 25 - 20 = 5$

Example 4. Solve this expression for $x = -2$. $12 + 3x$

Solution: Substitute −2 for x,
Then: $12 + 3x = 12 + 3(-2) = 12 - 6 = 6$

EVALUATING TWO VARIABLES

☑ To evaluate an algebraic expression, substitute a number for each variable.

☑ Perform the arithmetic operations to find the value of the expression.

Examples:

Example 1. Calculate this expression for a = 3 and $b = -2$. $3a - 6b$

Solution: First, substitute 3 for a, and -2 for b ,
Then: $3a - 6b = 3(3) - 6(-2)$
Now, use order of operation to find the answer: $3(3) - 6(-2) = 9 + 12 = 21$

Example 2. Evaluate this expression for x = 3 and $y = 1$. $3x + 5y$

Solution: Substitute 3 for x, and 1 for y ,
Then: $3x + 5y = 3(3) + 5(1) = 9 + 5 = 14$

Example 3. Find the value of this expression $5(3a - 2b)$ when $a = 1$ and $b = 2$.

Solution: Substitute 1 for a, and 2 for b ,
Then: $5(3a - 2b) = 15a - 10b = 15(1) - 10(2) = 15 - 20 = -5$

Example 4. Solve this expression. $4x - 3y$, $x = 3$, $y = 5$

Solution: Substitute 3 for x, and 5 for y and simplify.
Then: $4x - 3y = 4(3) - 3(5) = 12 - 15 = -3$

CHAPTER 6: PRACTICES

✎ Simplify each expression.

1) $(9x - 5x - 7 + 4) =$

2) $(-12x - 7x + 6 - 3) =$

3) $(24x - 10x - 3) =$

4) $(-10x + 23x - 6) =$

5) $(32x + 8 - 20x - 4) =$

6) $3 + 6x^2 - 6 =$

7) $3x + 6x^4 - 6x =$

8) $-1 - 2x^2 - 8 =$

9) $67 + 6x - 1 - 9 =$

10) $3x^2 + 9x - 11x - 2 =$

11) $-3x^2 - 5x - 7x + 6 - 7 =$

12) $9x - 2x^2 + 8x =$

13) $12x^2 + 6x - 3x^2 + 12 =$

14) $10x^2 - 8x - 5x^2 + 4 =$

✎ Simplify each polynomial.

15) $8x^2 + 2x^3 - 4x^2 + 10x =$ _____

16) $6x^4 + 3x^5 - 9x^4 + 7x^2 =$ _____

17) $10x^3 + 12x - 3x^2 - 7x^3 =$ _____

18) $(6x^3 - 2x^2) + (4x^2 - 14x) =$ _____

19) $(13x^4 + 5x^3) + (2x^3 - 6x^4) =$ _____

20) $(14x^5 - 9x^3) - (3x^3 + x^2) =$ _____

21) $(10x^4 + 6x^3) - (x^3 - 65) =$ _____

22) $(26x^4 + 5x^3) - (15x^3 - 3x^4) =$ _____

23) $(10x^2 + 8x^3) + (25x^2 + 4x^3) =$ _____

24) $(8x^4 - 3x^3) + (4x^3 - 7x^4) =$ _____

✑ **Use the distributive property to simply each expression.**

25) $3(6 + 9x) = $ _____   30) $(-x + 1)(-9) = $ _____

26) $6(4 - 3x) = $ _____   31) $(-3)(9x - 5) = $ _____

27) $(-8)(3 - 4x) = $ _____   32) $(2x + 10)6 = $ _____

28) $(2 - 5x)(-6) = $ _____   33) $(-1)(1 - 3x) = $ _____

29) $3(7 - 2x) = $ 34) $(6x - 1)(-9) = $ _____

✑ **Evaluate each expression using the value given.**

35) $x = -5$, $14 - x = $ ____   42) $x = 5$, $10 - x = $ ____

36) $x = -7$, $x + 10 = $ ____   43) $x = 2$, $28 - 5x = $ ____

37) $x = 4$, $6x - 8 = $ ____   44) $x = -9$, $100 - 6x = $ ____

38) $x = 3$, $9 - 3x = $ ____   45) $x = 10$, $50 - 8x = $ ____

39) $x = -8$, $6x - 9 = $ ____   46) $x = 3$, $61x - 3 = $ ____

40) $x = 7$, $18 - 3x = $ ____   47) $x = 3$, $25x - 2 = $ ____

41) $x = -1$, $14x - 3 = $ ____   48) $x = -1$, $13 - 4x = $ ____

✑ **Evaluate each expression using the values given.**

49) $x = 2, y = -1$, $3x - 6y = $ _____

50) $a = 3, b = 6$, $7a + 2b = $ _____

51) $x = 3, y = 2$, $5x - 23y + 9 = $ _____

52) $a = 7, b = 5$, $-9a + 3b + 8 = $ _____

53) $x = 3, y = 6$, $3x + 15 + 6y = $ _____

CHAPTER 6: ANSWERS

1) $4x - 3$

2) $-19x + 3$

3) $14x - 3$

4) $13x - 6$

5) $12x + 4$

6) $6x^2 - 3$

7) $6x^4 - 3x$

8) $-2x^2 - 9$

9) $6x + 57$

10) $3x^2 - 2x - 2$

11) $-3x^2 - 12x - 1$

12) $-2x^2 + 17x$

13) $9x^2 + 6x + 12$

14) $5x^2 - 8x + 4$

15) $2x^3 + 4x^2 + 10x$

16) $3x^5 - 3x^4 + 7x^2$

17) $3x^3 - 3x^2 + 12x$

18) $6x^3 + 2x^2 - 14x$

19) $7x^4 + 7x^3$

20) $14x^5 - 12x^3 - x^2$

21) $10x^4 + 5x^3 + 65$

22) $29x^4 - 10x^3$

23) $12x^3 + 35x^2$

24) $x^4 + x^3$

25) $27x + 18$

26) $-18x + 24$

27) $32x - 24$

28) $30x - 12$

29) $-6x + 21$

30) $9x - 9$

31) $-27x + 15$

32) $12x + 60$

33) $3x - 1$

34) $-54x + 9$

35) 19

36) 3

37) 16

38) 0

39) -57

40) -3

41) -17

42) 5

43) 18

44) 154

45) -30

46) 180

47) 73

48) 17

49) 12

50) 33

51) -22

52) -40

53) 60

CHAPTER 7:

EQUATIONS AND INEQUALITIES

Math Topics that you'll learn in this chapter:

▶ One-Step Equations

▶ Multi-Step Equations

▶ System of Equations

▶ Graphing Single–Variable Inequalities

▶ One-Step Inequalities

▶ Multi-Step Inequalities

ONE–STEP EQUATIONS

☑ The values of two expressions on both sides of an equation are equal. Example: $ax = b$. In this equation, ax is equal to b.

☑ Solving an equation means finding the value of the variable.

☑ You only need to perform one Math operation to solve the one-step equations.

☑ To solve a one-step equation, find the inverse (opposite) operation is being performed.

☑ The inverse operations are:

  ❖ Addition and subtraction

  ❖ Multiplication and division

Examples:

Example 1. Solve this equation for x. $3x = 18, x = ?$

Solution: Here, the operation is multiplication (variable x is multiplied by 3) and its inverse operation is division. To solve this equation, divide both sides of equation by 3: $3x = 18 \rightarrow \frac{3x}{3} = \frac{18}{3} \rightarrow x = 6$

Example 2. Solve this equation. $x + 15 = 0$, $x = ?$

Solution: In this equation 15 is added to the variable x. The inverse operation of addition is subtraction. To solve this equation, subtract 15 from both sides of the equation: $x + 15 - 15 = 0 - 15$. Then: $\rightarrow x = -15$

Example 3. Solve this equation for x. $x + 23 = 0$

Solution: Here, the operation is subtraction and its inverse operation is addition. To solve this equation, add 23 to both sides of the equation: $x + 23 - 23 = 0 - 23 \rightarrow x = -23$

MULTI–STEP EQUATIONS

☑ To solve a multi-step equation, combine "like" terms on one side.

☑ Bring variables to one side by adding or subtracting.

☑ Simplify using the inverse of addition or subtraction.

☑ Simplify further by using the inverse of multiplication or division.

☑ Check your solution by plugging the value of the variable into the original equation.

Examples:

Example 1. Solve this equation for x. $3x + 6 = 16 - 2x$

Solution: First, bring variables to one side by adding $2x$ to both sides. Then: $3x + 6 = 16 - 2x \rightarrow 3x + 6 + 2x = 16 - 2x + 2x$.
Simplify: $5x + 6 = 16$ Now, subtract 6 from both sides of the equation:
$5x + 6 - 6 = 16 - 6 \rightarrow 5x = 10 \rightarrow$ Divide both sides by 5:
$5x = 10 \rightarrow \dfrac{5x}{5} = \dfrac{10}{5} \rightarrow x = 2$
Let's check this solution by substituting the value of 2 for x in the original equation:
$x = 2 \rightarrow 3x + 6 = 16 - 2x \rightarrow 3(2) + 6 = 16 - 2(2) \rightarrow 6 + 6 = 16 - 4 \rightarrow 12 = 12$
The answer $x = 2$ is correct.

Example 2. Solve this equation for x. $-4x + 4 = 16$

Solution: Subtract 4 from both sides of the equation.
$-4x + 4 - 4 = 16 - 4 \rightarrow -4x = 12$
Divide both sides by -4, then: $-4x = 12 \rightarrow \dfrac{-4x}{-4} = \dfrac{12}{-4} \rightarrow x = -3$
Now, check the solution:
$x = -3 \rightarrow -4x + 4 = 16 \rightarrow -4(-3) + 4 = 16 \rightarrow 16 = 16$
The answer $x = -2$ is correct.

SYSTEM OF EQUATIONS

☑ A system of equations contains two equations and two variables. For example, consider the system of equations: $x - y = 1, x + y = 5$

☑ The easiest way to solve a system of equations is using the elimination method. The elimination method uses the addition property of equality. You can add the same value to each side of an equation.

☑ For the first equation above, you can add $x + y$ to the left side and 5 to the right side of the first equation: $x - y + (x + y) = 1 + 5$. Now, if you simplify, you get: $x - y + (x + y) = 1 + 5 \rightarrow 2x = 6 \rightarrow x = 3$. Now, substitute 3 for the x in the first equation: $3 - y = 1$. By solving this equation, $y = 2$

Example:

What is the value of x + y in this system of equations?

$$\begin{cases} x + 2y = 6 \\ 2x - y = -8 \end{cases}$$

Solution: Solving a System of Equations by Elimination:
Multiply the first equation by (-2), then add it to the second equation.

$$\begin{array}{l} -2(x + 2y = 6) \\ \underline{2x - y = -8} \end{array} \Rightarrow \begin{array}{l} -2x - 4y = -12 \\ \underline{2x - y = -8} \end{array} \Rightarrow -5y = -20 \Rightarrow y = 4$$

Plug in the value of y into one of the equations and solve for x.
$x + 2(4) = 6 \Rightarrow x + 8 = 6 \Rightarrow x = 6 - 8 \Rightarrow x = -2$
Thus, $x + y = -2 + 4 = 2$

GRAPHING SINGLE–VARIABLE INEQUALITIES

☑ An inequality compares two expressions using an inequality sign.

☑ Inequality signs are: "less than" <, "greater than" >, "less than or equal to" ≤, and "greater than or equal to" ≥.

☑ To graph a single–variable inequality, find the value of the inequality on the number line.

☑ For less than (<) or greater than (>) draw an open circle on the value of the variable. If there is an equal sign too, then use a filled circle.

☑ Draw an arrow to the right for greater or to the left for less than.

Examples:

Example 1. Draw a graph for this inequality. $x > 3$

Solution: Since the variable is greater than 3, then we need to find 3 in the number line and draw an open circle on it. Then, draw an arrow to the right.

Example 2. Graph this inequality. $x \leq -4$.

Solution: Since the variable is less than or equal to −4, then we need to find −4 in the number line and draw a filled circle on it. Then, draw an arrow to the left.

ONE–STEP INEQUALITIES

☑ An inequality compares two expressions using an inequality sign.

☑ Inequality signs are: "less than" <, "greater than" >, "less than or equal to" ≤, and "greater than or equal to" ≥.

☑ You only need to perform one Math operation to solve the one-step inequalities.

☑ To solve one-step inequalities, find the inverse (opposite) operation is being performed.

☑ For dividing or multiplying both sides by negative numbers, flip the direction of the inequality sign.

Examples:

Example 1. Solve this inequality for x. $x + 3 \geq 4$

Solution: The inverse (opposite) operation of addition is subtraction. In this inequality, 3 is added to x. To isolate x we need to subtract 3 from both sides of the inequality.

Then: $x + 3 \geq 4 \rightarrow x + 3 - 3 \geq 4 - 3 \rightarrow x \geq 1$. The solution is: $x \geq 1$

Example 2. Solve the inequality. $x - 5 > -4$.

Solution: 5 is subtracted from x. Add 5 to both sides.

$x - 5 > -4 \rightarrow x - 5 + 5 > -4 + 5 \rightarrow x > 1$

Example 3. Solve. $2x \leq -4$.

Solution: 2 is multiplied to x. Divide both sides by 2.

Then: $2x \leq -4 \rightarrow \frac{2x}{2} \leq \frac{-4}{2} \rightarrow x \leq -2$

Example 4. Solve. $-6x \leq 12$.

Solution: -6 is multiplied to x. Divide both sides by -6. Remember when dividing or multiplying both sides of an inequality by negative numbers, flip the direction of the inequality sign.

Then: $-6x \leq 12 \rightarrow \frac{-6x}{-6} \geq \frac{12}{-6} \rightarrow x \geq -2$

MULTI–STEP INEQUALITIES

☑ To solve a multi-step inequality, combine "like" terms on one side.

☑ Bring variables to one side by adding or subtracting.

☑ Isolate the variable.

☑ Simplify using the inverse of addition or subtraction.

☑ Simplify further by using the inverse of multiplication or division.

☑ For dividing or multiplying both sides by negative numbers, flip the direction of the inequality sign.

Examples:

Example 1. Solve this inequality. $2x - 3 \leq 5$

Solution: In this inequality, 3 is subtracted from $2x$. The inverse of subtraction is addition. Add 3 to both sides of the inequality:
$2x - 3 + 3 \leq 5 + 3 \rightarrow 2x \leq 8$
Now, divide both sides by 2. Then: $2x \leq 8 \rightarrow \frac{2x}{2} \leq \frac{8}{2} \rightarrow x \leq 4$
The solution of this inequality is $x \leq 4$.

Example 2. Solve this inequality. $3x + 9 < 12$

Solution: First, subtract 9 from both sides: $3x + 9 - 9 < 12 - 9$
Then simplify: $3x + 9 - 9 < 12 - 9 \rightarrow 3x < 3$
Now divide both sides by 3: $\frac{3x}{3} < \frac{3}{3} \rightarrow x < 1$

Example 3. Solve this inequality. $-2x + 4 \geq 6$

Solution: First, subtract 4 from both sides:
$-2x + 4 - 4 \geq 6 - 4 \rightarrow -2x \geq 2$
Divide both sides by -2. Remember that you need to flip the direction of inequality sign. $-2x \geq 2 \rightarrow \frac{-2x}{-2} \leq \frac{2}{-2} \rightarrow x \leq -1$

CHAPTER 7: PRACTICES

✍ Solve each equation. (One–Step Equations)

1) $x + 8 = 4, x =$ _____

2) $3 = 12 - x, x =$ _____

3) $-4 = 9 + x, x =$ _____

4) $x - 6 = -9, x =$ _____

5) $18 = x + 8, x =$ _____

6) $15 - x = -4, x =$ _____

7) $25 - x = 8, x =$ _____

8) $6 + x = 27, x =$ _____

9) $10 - x = -8, x =$ _____

10) $36 - x = -5, x =$ _____

✍ Solve each equation. (Multi–Step Equations)

11) $6(x + 8) = 24, \ x =$ _____

12) $-9(9 - x) = 18, x =$ _____

13) $7 = -7(x + 3), x =$ _____

14) $-16 = 2(10 - 6x), x =$ _____

15) $6(x + 1) = -24, x =$ _____

16) $-3(7 + 9x) = 33, x =$ _____

17) $-7(5 - x) = 14, x =$ _____

18) $-1(3 - x) = 10, x =$ _____

✍ Solve each system of equations.

19) $\begin{cases} -2x + 2y = -4 & x = \\ 4x - 9y = 28 & y = \end{cases}$

20) $\begin{cases} x + 8y = -5 & x = \\ 2x + 6y = 0 & y = \end{cases}$

21) $\begin{cases} 4x - 3y = -2 & x = \\ x - y = 3 & y = \end{cases}$

22) $\begin{cases} 2x + 9y = 17 & x = \\ -3x + 8y = 39 & y = \end{cases}$

✎ **Draw a graph for each inequality.**

23) $x \leq -3$

24) $x > -5$

✎ **Solve each inequality and graph it.**

25) $x - 2 \geq -2$

26) $2x - 3 < 9$

✎ **Solve each inequality.**

27) $4x + 12 > -8$ 35) $12x + 8 < 32$

28) $3x + 14 > 5$ 36) $8(4 + x) \geq 16$

29) $-16 + 3x \leq 20$ 37) $2(x - 5) \geq 18$

30) $-18 + 6x \leq -24$ 38) $x + 10 < 3$

31) $8 + 2x \leq 16$ 39) $2(x - 4) \geq 20$

32) $5(x + 2) \geq 6$ 40) $-8 + 9x > 28$

33) $2(3 + x) \geq 10$ 41) $-4 + 8x > 60$

34) $6x - 10 < 14$ 42) $-2 + 7x > 40$

CHAPTER 7: ANSWERS

1) -4

2) 9

3) -13

4) -3

5) 10

6) 19

7) 17

8) 21

9) 18

10) 41

11) -4

12) 11

13) -4

14) 3

15) -5

16) -2

17) 7

18) 13

19) $x = -2, y = -4$

20) $x = 3, y = -1$

21) $x = -11, y = -14$

22) $x = -5, y = 3$

23) $x \leq -3$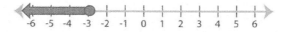

24) $x > -5$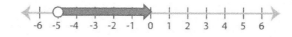

25) $x \geq 0$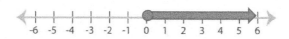

26) $x < 6$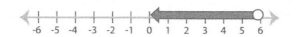

27) $x > -5$

28) $x > -3$

29) $x \leq 12$

30) $x \leq -1$

31) $x \leq 4$

32) $x \geq -\frac{4}{5}$

33) $x \geq 2$

34) $x < 4$

35) $x < 2$

36) $x \geq -2$

37) $x \geq 14$

38) $x < -7$

39) $x \geq 14$

40) $x > 4$

41) $x > 8$

42) $x > 6$

CHAPTER 8:

LINES AND SLOPE

Math Topics that you'll learn in this chapter:

▶ Finding Slope

▶ Graphing Lines Using Slope–Intercept Form

▶ Writing Linear Equations

▶ Graphing Linear Inequalities

▶ Finding Midpoint

▶ Finding Distance of Two Points

FINDING SLOPE

- ☑ The slope of a line represents the direction of a line on the coordinate plane.

- ☑ A coordinate plane contains two perpendicular number lines. The horizontal line is x and the vertical line is y. The point at which the two axes intersect is called the origin. An ordered pair (x, y) shows the location of a point.

- ☑ A line on a coordinate plane can be drawn by connecting two points.

- ☑ To find the slope of a line, we need the equation of the line or two points on the line.

- ☑ The slope of a line with two points A (x_1, y_1) and B (x_2, y_2) can be found by using this formula: $\frac{y_2 - y_1}{x_2 - x_1} = \frac{rise}{run}$

- ☑ The equation of a line is typically written as $y = mx + b$ where m is the slope and b is the y-intercept.

Examples:

Example 1. Find the slope of the line through these two points:

A$(2, -7)$ and $B(4, 3)$.

Solution: Slope $= \frac{y_2 - y_1}{x_2 - x_1}$. Let (x_1, y_1) be A$(2, -7)$ and (x_2, y_2) be $B(4, 3)$.

(Remember, you can choose any point for (x_1, y_1) and (x_2, y_2)).

Then: slope $= \frac{y_2 - y_1}{x_2 - x_1} = \frac{3 - 7}{4 - 2} = \frac{10}{2} = 5$

The slope of the line through these two points is 5.

Example 2. Find the slope of the line with equation $y = 3x + 6$

Solution: when the equation of a line is written in the form of $y = mx + b$, the slope is m. In this line: $y = 3x + 6$, the slope is 3.

GRAPHING LINES USING SLOPE–INTERCEPT FORM

☑ Slope–intercept form of a line: given the slope **m** and the **y**-intercept (the intersection of the line and y-axis) **b**, then the equation of the line is:

$$y = mx + b$$

☑ To draw the graph of a linear equation in a slope-intercept form on the xy coordinate plane, find two points on the line by plugging two values for x and calculating the values of y.

☑ You can also use the slope (m) and one point to graph the line.

Example:

Example 1. Sketch the graph of $y = 2x - 4$.

Solution: To graph this line, we need to find two points. When x is zero the value of y is -4. And when x is 2 the value of y is 0.

$$x = 0 \rightarrow y = 2(0) - 4 = -4,$$
$$y = 0 \rightarrow 0 = 2x - 4 \rightarrow x = 2$$

Now, we have two points: $(0, -4)$ and $(2, 0)$.
Find the points on the coordinate plane and graph the line. Remember that the slope of the line is 2.

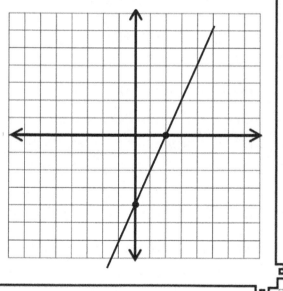

WRITING LINEAR EQUATIONS

☑ The equation of a line in slope-intercept form: $y = mx + b$

☑ To write the equation of a line, first identify the slope.

☑ Find the y-intercept. This can be done by substituting the slope and the coordinates of a point (x, y) on the line.

Examples:

Example 1. What is the equation of the line that passes through $(2, -4)$ and has a slope of 8?

Solution: The general slope-intercept form of the equation of a line is $y = mx + b$, where m is the slope and b is the y-intercept.
By substitution of the given point and given slope:
$y = mx + b \rightarrow -4 = (2)(8) + b$. So, $b = -4 - 16 = -20$, and the required equation is $y = 8x - 20$

Example 2. Write the equation of the line through two points $A(2, 1)$ and $B(-2, 5)$.

Solution: First, find the slope: $Slop = \frac{y_2 - y_1}{x_2 - x_1} = \frac{5 - 1}{-2 - 2} = \frac{4}{-4} = -1 \rightarrow m = -1$
To find the value of b, use either points and plug in the values of x and y in the equation. The answer will be the same: $y = -x + b$. Let's check both points. Then: $(2, 1) \rightarrow y = mx + b \rightarrow 1 = -1(2) + b \rightarrow b = 3$
$(-2, 5) \rightarrow y = mx + b \rightarrow 5 = -1(-2) + b \rightarrow b = 3$.
The y-intercept of the line is 3. The equation of the line is: $y = -x + 3$

Example 3. What is the equation of the line that passes through $(2, -1)$ and has a slope of 5?

Solution: The general slope-intercept form of the equation of a line is $y = mx + b$, where m is the slope and b is the y-intercept. By substitution of the given point and given slope: $y = mx + b \rightarrow -1 = (5)(2) + b$
So, $b = -1 - 10 = -11$, and the equation of the line is: $y = 5x - 11$.

FINDING MIDPOINT

☑ The middle of a line segment is its midpoint.

☑ The Midpoint of two endpoints A (x_1, y_1) and B (x_2, y_2) can be found using this formula: M $(\frac{x_1+x_2}{2}, \frac{y_1+y_2}{2})$

Examples:

Example 1. Find the midpoint of the line segment with the given endpoints. $(1, -3), (3, 7)$

Solution: Midpoint $= \left(\frac{x_1+x_2}{2}, \frac{y_1+y_2}{2}\right) \rightarrow (x_1, y_1) = (1, -3)$ and $(x_2, y_2) = (3, 7)$
Midpoint $= \left(\frac{1+3}{2}, \frac{-3+7}{2}\right) \rightarrow \left(\frac{4}{2}, \frac{4}{2}\right) \rightarrow M(2, 2)$

Example 2. Find the midpoint of the line segment with the given endpoints. $(-4, 5), (8, -7)$

Solution: Midpoint $= \left(\frac{x_1+x_2}{2}, \frac{y_1+y_2}{2}\right) \rightarrow (x_1, y_1) = (-4, 5)$ and $(x_2, y_2) = (8, -7)$
Midpoint $= \left(\frac{-4+8}{2}, \frac{5-7}{2}\right) \rightarrow \left(\frac{4}{2}, \frac{-2}{2}\right) \rightarrow M(2, -1)$

Example 3. Find the midpoint of the line segment with the given endpoints. $(5, -2), (1, 10)$

Solution: Midpoint $= \left(\frac{x_1+x_2}{2}, \frac{y_1+y_2}{2}\right) \rightarrow (x_1, y_1) = (5, -2)$ and $(x_2, y_2) = (1, 10)$
Midpoint $= \left(\frac{5+1}{2}, \frac{-2+10}{2}\right) \rightarrow \left(\frac{6}{2}, \frac{8}{2}\right) \rightarrow M(3, 4)$

Example 4. Find the midpoint of the line segment with the given endpoints. $(2, 3), (12, -9)$

Solution: Midpoint $= \left(\frac{x_1+x_2}{2}, \frac{y_1+y_2}{2}\right) \rightarrow (x_1, y_1) = (2, 3)$ and $(x_2, y_2) = (12, -3)$
Midpoint $= \left(\frac{2+12}{2}, \frac{3-9}{2}\right) \rightarrow \left(\frac{14}{2}, \frac{-6}{2}\right) \rightarrow M(7, -3)$

FINDING DISTANCE OF TWO POINTS

⌾ Use the following formula to find the distance of two points with the coordinates A (x_1, y_1) and B (x_2, y_2):

$$d = \sqrt{(x_2 - x_1)^2 + (y_2 - y_1)^2}$$

Examples:

Example 1. Find the distance between $(4, 6)$ and $(1, 2)$.

Solution: Use distance of two points formula: $d = \sqrt{(x_2 - x_1)^2 + (y_2 - y_1)^2}$ $(x_1, y_1) = (4, 6)$ and $(x_2, y_2) = (1, 2)$. Then: $d = \sqrt{(x_2 - x_1)^2 + (y_2 - y_1)^2} \rightarrow$

$$d = \sqrt{(1 - (4))^2 + (2 - 6)^2} = \sqrt{(-3)^2 + (-4)^2} = \sqrt{9 + 16} = \sqrt{25} = 5 \rightarrow d = 5$$

Example 2. Find the distance of two points $(-6, -10)$ and $(-2, -10)$.

Solution: Use distance of two points formula: $d = \sqrt{(x_2 - x_1)^2 + (y_2 - y_1)^2}$ $(x_1, y_1) = (-6, -10)$, and $(x_2, y_2) = (-2, -10)$

Then: $d = \sqrt{(x_2 - x_1)^2 + (y_2 - y_1)^2} \rightarrow d = \sqrt{(-2 - (-6))^2 + (-10 - (-10))^2} =$

$\sqrt{(4)^2 + (0)^2} = \sqrt{16 + 0} = \sqrt{16} = 4$. Then: $d = 4$

Example 3. Find the distance between $(-6, 5)$ and $(-2, 2)$.

Solution: Use distance of two points formula: $d = \sqrt{(x_2 - x_1)^2 + (y_2 - y_1)^2}$ $(x_1, y_1) = (-6, 5)$ and $(x_2, y_2) = (-2, 2)$. Then: $d = \sqrt{(x_2 - x_1)^2 + (y_2 - y_1)^2}$

$$d = \sqrt{(-2 - (-6))^2 + (2 - 5)^2} = \sqrt{(4)^2 + (-3)^2} = \sqrt{16 + 9} = \sqrt{25} = 5$$

CHAPTER 8: PRACTICES

✍ Find the slope of each line.

1) $y = x - 3$

2) $y = -6x + 4$

3) $y = 3x - 9$

4) Line through $(-1, 3)$ and $(5, 0)$

5) Line through $(4, 0)$ and $(-2, 6)$

6) Line through $(-3, -6)$ and $(0, 3)$

✍ Sketch the graph of each line. (Using Slope–Intercept Form)

7) $y = x + 4$

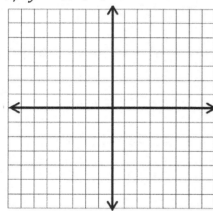

8) $y = 2x - 5$

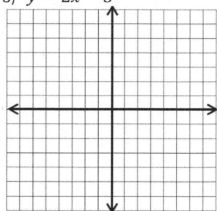

✍ Solve.

9) What is the equation of a line with slope 3 and intercept 18? _____

10) What is the equation of a line with slope 2 and passes through point $(2, 6)$?

_____

11) What is the equation of a line with slope -4 and passes through point $(-4, 8)$?

_____

12) The slope of a line is -2 and it passes through point $(-4, 3)$. What is the equation of the line? _____

13) The slope of a line is 5 and it passes through point $(-6, 3)$. What is the equation of the line? _____

✍ **Sketch the graph of each linear inequality.**

14) $y > 2x - 2$

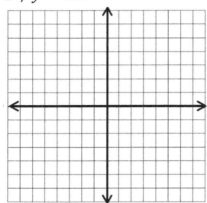

15) $y < -x + 3$

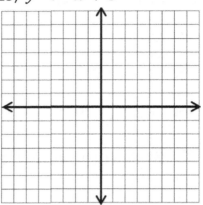

✍ **Find the midpoint of the line segment with the given endpoints.**

16) $(3, 6), (-1, 8)$

17) $(-2, 4), (8, 4)$

18) $(8, -3), (-2, 1)$

19) $(15, -9), (-3, 1)$

20) $(-10, 4), (6, 8)$

21) $(6, 12), (2, -4)$

22) $(4, 8), (-2, 0)$

23) $(0, 8), (-6, 6)$

✍ **Find the distance between each pair of points.**

24) $(-1, 6), (-5, 3)$

25) $(2, -2), (7, 10)$

26) $(-1, -4), (5, 4)$

27) $(6, -1), (-6, 8)$

28) $(2, -5), (-6, 10)$

29) $(0, 6), (4, 6)$

30) $(6, 3), (9, -1)$

31) $(0, -2), (10, 22)$

32) $(5, -6), (-11, 24)$

33) $(6, -10), (-6, 6)$

CHAPTER 8: ANSWERS

1) 1

2) −6

3) 3

4) $-\frac{1}{2}$

5) −1

6) 3

7) $y = x + 4$

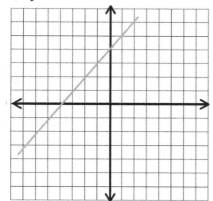

8) $y = 2x - 5$

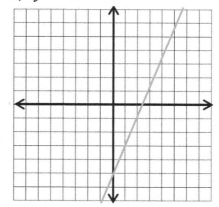

9) $y = 3x + 18$

10) $y = 2x + 2$

11) $y = -4x - 8$

12) $y = -2x - 5$

13) $y = 5x + 33$

14) $y > 2x - 2$

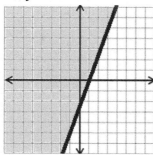

15) $y < -x + 3$

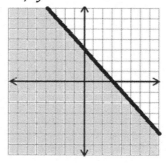

16) $(1, 7)$

17) $(3, 4)$

18) $(3, -1)$

19) $(6, -4)$

20) $(-2, 6)$

21) $(4, 4)$

22) $(1, 4)$

23) $(-3, 7)$

24) 5

25) 13

26) 10

27) 15

28) 17

29) 4

30) 5

31) 26

32) 34

33) 20

CHAPTER 9:

EXPONENTS AND VARIABLES

Math Topics that you'll learn in this chapter:

▶ Multiplication Property of Exponents

▶ Division Property of Exponents

▶ Powers of Products and Quotients

▶ Zero and Negative Exponents

▶ Negative Exponents and Negative Bases

▶ Scientific Notation

▶ Radicals

MULTIPLICATION PROPERTY OF EXPONENTS

☑ Exponents are shorthand for repeated multiplication of the same number by itself. For example, instead of 2×2, we can write 2^2. For $3 \times 3 \times 3 \times 3$, we can write 3^4

☑ In algebra, a variable is a letter used to stand for a number. The most common letters are: x, y, z, a, b, c, m, and n.

☑ Exponent's rules: $x^a \times x^b = x^{a+b}$, $\frac{x^a}{x^b} = x^{a-b}$

$$(x^a)^b = x^{a \times b} \qquad\qquad (xy)^a = x^a \times y^a \qquad\qquad \left(\frac{a}{b}\right)^c = \frac{a^c}{b^c}$$

Examples:

Example 1. Multiply. $4x^3 \times 2x^2$

Solution: Use Exponent's rules: $x^a \times x^b = x^{a+b} \rightarrow x^3 \times x^2 = x^{3+2} = x^5$
Then: $4x^3 \times 2x^2 = 8x^5$

Example 2. Simplify. $\left(x^3y^5\right)^2$

Solution: Use Exponent's rules: $(x^a)^b = x^{a \times b}$.
Then: $\left(x^3y^5\right)^2 = x^{3 \times 2}y^{5 \times 2} = x^6y^{10}$

Example 3. Multiply. $-2x^5 \times 7x^3$

Solution: Use Exponent's rules: $x^a \times x^b = x^{a+b} \rightarrow x^5 \times x^3 = x^{5+3} = x^8$
Then: $-2x^5 \times 7x^3 = -14x^8$

Example 4. Simplify. $(x^2y^4)^3$

Solution: Use Exponent's rules: $(x^a)^b = x^{a \times b}$.
Then: $(x^2y^4)^3 = x^{2 \times 3}y^{4 \times 3} = x^6y^{12}$

DIVISION PROPERTY OF EXPONENTS

☑ Exponents are shorthand for repeated multiplication of the same number by itself. For example, instead of 3×3, we can write 3^2. For $2 \times 2 \times 2$, we can write 2^3

☑ For division of exponents use following formulas:

$$\frac{x^a}{x^b} = x^{a-b} \ , \ x \neq 0, \quad \frac{x^a}{x^b} = \frac{1}{x^{b-a}} \ , \ x \neq 0, \qquad \frac{1}{x^b} = x^{-b}$$

Examples:

Example 1. Simplify. $\frac{12x^2y}{4xy^3} =$

Solution: First, cancel the common factor: $4 \rightarrow \frac{12x^2y}{4xy^3} = \frac{3x^2y}{xy^3}$

Use Exponent's rules: $\frac{x^a}{x^b} = x^{a-b} \ \rightarrow \frac{x^2}{x} = x^{2-1} = x$ and $\frac{y}{y^3} = \frac{1}{y^{3-1}} = \frac{1}{y^2}$

Then: $\frac{12x^2y}{4xy^3} = \frac{3x}{y^2}$

Example 2. Simplify. $\frac{18x^6}{2x^3} =$

Solution: Use Exponent's rules: $\frac{x^a}{x^b} = x^{b-a} \ \rightarrow \frac{x^6}{x^3} = x^{6-3} = x^3$

Then: $\frac{18x^6}{2x^3} = 9x^3$

Example 3. Simplify. $\frac{8x^3y}{40x^2y^3} =$

Solution: First, cancel the common factor: $8 \rightarrow \frac{8x^3y}{40x^2y^3} = \frac{x^3y}{5x^2y^3}$

Use Exponent's rules: $\frac{x^a}{x^b} = x^{a-b} \ \rightarrow \frac{x^3}{x^2} = x^{3-2} = x$

Then: $\frac{8x^3y}{40x^2y^3} = \frac{xy}{5y^3} \rightarrow$ now cancel the common factor: $y \rightarrow \frac{xy}{5y^3} = \frac{x}{5y^2}$

POWERS OF PRODUCTS AND QUOTIENTS

☑ Exponents are shorthand for repeated multiplication of the same number by itself. For example, instead of $2 \times 2 \times 2$, we can write 2^3. For $3 \times 3 \times 3 \times 3$, we can write 3^4

☑ For any nonzero numbers a and b and any integer x, $(ab)^x = a^x \times b^x$ and $\left(\frac{a}{b}\right)^c = \frac{a^c}{b^c}$

Examples:

Example 1. Simplify. $(6x^2y^4)^2$

Solution: Use Exponent's rules: $(x^a)^b = x^{a \times b}$
$(6x^2y^4)^2 = (6)^2(x^2)^2(y^4)^2 = 36x^{2 \times 2}y^{4 \times 2} = 36x^4y^8$

Example 2. Simplify. $\left(\frac{5x}{2x^2}\right)^2$

Solution: First, cancel the common factor: $x \to \left(\frac{5x}{2x^2}\right)^2 = \left(\frac{5}{2x}\right)^2$
Use Exponent's rules: $\left(\frac{a}{b}\right)^c = \frac{a^c}{b^c}$, Then: $\left(\frac{5}{2x}\right)^2 = \frac{5^2}{(2x)^2} = \frac{25}{4x^2}$

Example 3. Simplify. $\left(3x^5y^4\right)^2$

Solution: Use Exponent's rules: $(x^a)^b = x^{a \times b}$
$$\left(3x^5y^4\right)^2 = (3)^2\left(x^5\right)^2(y^4)^2 = 9x^{5 \times 2}y^{4 \times 2} = 9x^{10}y^8$$

Example 4. Simplify. $\left(\frac{2x}{3x^2}\right)^2$

Solution: First, cancel the common factor: $x \to \left(\frac{2x}{3x^2}\right)^2 = \left(\frac{2}{3x}\right)^2$
Use Exponent's rules: $\left(\frac{a}{b}\right)^c = \frac{a^c}{b^c}$, Then: $\left(\frac{2}{3x}\right)^2 = \frac{2^2}{(3x)^2} = \frac{4}{9x^2}$

ZERO AND NEGATIVE EXPONENTS

☑ Zero-Exponent Rule: $a^0 = 1$, this means that anything raised to the zero power is 1. For example: $(5xy)^0 = 1$

☑ A negative exponent simply means that the base is on the wrong side of the fraction line, so you need to flip the base to the other side. For instance, "x^{-2}" (pronounced as "ecks to the minus two") just means "x^2" but underneath, as in $\frac{1}{x^2}$.

Examples:

Example 1. Evaluate. $\left(\frac{2}{3}\right)^{-2} =$

Solution: Use negative exponent's rule: $\left(\frac{x^a}{x^b}\right)^{-2} = \left(\frac{x^b}{x^a}\right)^2 \rightarrow \left(\frac{2}{3}\right)^{-2} = \left(\frac{3}{2}\right)^2 =$
Then: $\left(\frac{3}{2}\right)^2 = \frac{3^2}{2^2} = \frac{9}{4}$

Example 2. Evaluate. $\left(\frac{4}{5}\right)^{-3} =$

Solution: Use negative exponent's rule: $\left(\frac{x^a}{x^b}\right)^{-2} = \left(\frac{x^b}{x^a}\right)^2 \rightarrow \left(\frac{4}{5}\right)^{-3} = \left(\frac{5}{4}\right)^3 =$
Then: $\left(\frac{5}{4}\right)^3 = \frac{5^3}{4^3} = \frac{125}{64}$

Example 3. Evaluate. $\left(\frac{x}{y}\right)^0 =$

Solution: Use zero-exponent Rule: $a^0 = 1$
Then: $\left(\frac{x}{y}\right)^0 = 1$

Example 4. Evaluate. $\left(\frac{5}{6}\right)^{-1} =$

Solution: Use negative exponent's rule: $\left(\frac{x^a}{x^b}\right)^{-2} = \left(\frac{x^b}{x^a}\right)^2 \rightarrow \left(\frac{5}{6}\right)^{-1} = \left(\frac{6}{5}\right)^1 = \frac{6}{5}$

NEGATIVE EXPONENTS AND NEGATIVE BASES

☑ A negative exponent is the reciprocal of that number with a positive exponent. $(3)^{-2} = \frac{1}{3^2}$

☑ To simplify a negative exponent, make the power positive!

☑ The parenthesis is important! -5^{-2} is not the same as $(-5)^{-2}$

$$-5^{-2} = -\frac{1}{5^2} \text{ and } (-5)^{-2} = +\frac{1}{5^2}$$

Examples:

Example 1. Simplify. $\left(\frac{5a}{6c}\right)^{-2} =$

Solution: Use negative exponent's rule: $\left(\frac{x^a}{x^b}\right)^{-2} = \left(\frac{x^b}{x^a}\right)^2 \rightarrow \left(\frac{5a}{6c}\right)^{-2} = \left(\frac{6c}{5a}\right)^2$

Now use exponent's rule: $\left(\frac{a}{b}\right)^c = \frac{a^c}{b^c} \rightarrow = \left(\frac{6c}{5a}\right)^2 = \frac{6^2 c^2}{5^2 a^2}$

Then: $\frac{6^2 c^2}{5^2 a^2} = \frac{36c^2}{25a^2}$

Example 2. Simplify. $\left(\frac{2x}{3yz}\right)^{-3} =$

Solution: Use negative exponent's rule: $\left(\frac{x^a}{x^b}\right)^{-2} = \left(\frac{x^b}{x^a}\right)^2 \rightarrow \left(\frac{2x}{3yz}\right)^{-3} = \left(\frac{3yz}{2x}\right)^3$

Now use exponent's rule: $\left(\frac{a}{b}\right)^c = \frac{a^c}{b^c} \rightarrow \left(\frac{3yz}{2x}\right)^3 = \frac{3^3 y^3 z^3}{2^3 x^3} = \frac{27y^3 z^3}{8x^3}$

Example 3. Simplify. $\left(\frac{3a}{2c}\right)^{-2} =$

Solution: Use negative exponent's rule: $\left(\frac{x^a}{x^b}\right)^{-2} = \left(\frac{x^b}{x^a}\right)^2 \rightarrow \left(\frac{3a}{2c}\right)^{-2} = \left(\frac{2c}{3a}\right)^2$

Now use exponent's rule: $\left(\frac{a}{b}\right)^c = \frac{a^c}{b^c} \rightarrow = \left(\frac{2c}{3a}\right)^2 = \frac{2^2 c^2}{3^2 a^2}$

Then: $\frac{2^2 c^2}{3^2 a^2} = \frac{4c^2}{9a^2}$

SCIENTIFIC NOTATION

☑ Scientific notation is used to write very big or very small numbers in decimal form.

☑ In scientific notation, all numbers are written in the form of: $m \times 10^n$, where m is greater than 1 and less than 10.

☑ To convert a number from scientific notation to standard form, move the decimal point to the left (if the exponent of ten is a negative number), or to the right (if the exponent is positive).

Examples:

Example 1. Write 0.00015 in scientific notation.

Solution: First, move the decimal point to the right so you have a number between 1 and 10. That number is 1.5. Now, determine how many places the decimal moved in step 1 by the power of 10. We moved the decimal point 4 digits to the right. Then: $10^{-4} \rightarrow$ When the decimal moved to the right, the exponent is negative. Then: $0.00015 = 1.5 \times 10^{-4}$

Example 2. Write 9.5×10^{-5} in standard notation.

Solution: $10^{-5} \rightarrow$ When the decimal moved to the right, the exponent is negative. Then: $9.5 \times 10^{-5} = 0.000095$

Example 3. Write 0.00012 in scientific notation.

Solution: First, move the decimal point to the right so you have a number between 1 and 10. Then: $m = 1.2$, Now, determine how many places the decimal moved in step 1 by the power of 10.
$10^{-4} \rightarrow$ Then: $0.00012 = 1.2 \times 10^{-4}$

Example 4. Write 8.3×10^5 in standard notation.
Solution: $10^{-5} \rightarrow$ The exponent is positive 5. Then, move the decimal point to the right five digits. (remember $8.3 = 8.30000$),
Then: $8.3 \times 10^5 = 830000$

RADICALS

✓ If n is a positive integer and x is a real number, then: $\sqrt[n]{x} = x^{\frac{1}{n}}$,

$$\sqrt[n]{xy} = x^{\frac{1}{n}} \times y^{\frac{1}{n}}, \ \sqrt[n]{\frac{x}{y}} = \frac{x^{\frac{1}{n}}}{y^{\frac{1}{n}}}, \text{ and } \sqrt[n]{x} \times \sqrt[n]{y} = \sqrt[n]{xy}$$

✓ A square root of x is a number r whose square is: $r^2 = x$ (r is a square root of x)

✓ To add and subtract radicals, we need to have the same values under the radical. For example: $\sqrt{3} + \sqrt{3} = 2\sqrt{3}$, $3\sqrt{5} - \sqrt{5} = 2\sqrt{5}$

Examples:

Example 1. Find the square root of $\sqrt{169}$.

Solution: First, factor the number: $169 = 13^2$, Then: $\sqrt{169} = \sqrt{13^2}$, Now use radical rule: $\sqrt[n]{a^n} = a$. Then: $\sqrt{169} = \sqrt{13^2} = 13$

Example 2. Evaluate. $\sqrt{9} \times \sqrt{25} =$

Solution: Find the values of $\sqrt{9}$ and $\sqrt{25}$. Then: $\sqrt{9} \times \sqrt{25} = 3 \times 5 = 15$

Example 3. Solve. $7\sqrt{2} + 4\sqrt{2}$.

Solution: Since we have the same values under the radical, we can add these two radicals: $7\sqrt{2} + 4\sqrt{2} = 11\sqrt{2}$

Example 4. Evaluate. $\sqrt{2} \times \sqrt{8} =$

Solution: Use this radical rule: $\sqrt[n]{x} \times \sqrt[n]{y} = \sqrt[n]{xy} \rightarrow \sqrt{2} \times \sqrt{8} = \sqrt{16}$ The square root of 16 is 4. Then: $\sqrt{2} \times \sqrt{8} = \sqrt{16} = 4$

CHAPTER 9: PRACTICES

✍ Find the products.

1) $2x^3 \times 4xy^2 =$

2) $6x^2y \times 8x^2y^2 =$

3) $5x^3y^2 \times 3x^2y^3 =$

4) $7xy^4 \times 4x^2y =$

5) $3x^4y^5 \times 9x^3y^2 =$

6) $6x^3y^2 \times 7x^3y^3 =$

7) $4x^3y^6 \times 2x^4y^2 =$

8) $7x^4y^3 \times 3x^3y^2 =$

9) $10x^5y^2 \times 10x^4y^3 =$

10) $8x^2y^3 \times 5x^6y^2 =$

11) $9y^5 \times 2x^6y^3 =$

12) $7x^4 \times 7x^2y^2 =$

✍ Simplify.

13) $\frac{3^3 \times 3^4}{3^9 \times 3} =$

14) $\frac{6x}{30x^2} =$

15) $\frac{18x^4}{6x^3} =$

16) $\frac{42x^3}{56\ ^3y^2} =$

17) $\frac{18y^3}{54x^4y^4} =$

18) $\frac{150x^3y^5}{50x^2y^3} =$

19) $\frac{2^3 \times 2^2}{7^2 \times 7} =$

20) $\frac{12x}{2x^2} =$

21) $\frac{25x^6}{5x^3} =$

22) $\frac{48y^4}{56x^5y^3} =$

✍ Solve.

23) $(3x^2y^6)^3 =$

24) $(2x^3y^4)^5 =$

25) $(2x \times 5xy^2)^2 =$

26) $(3x \times 2y^3)^2 =$

27) $\left(\frac{8x}{x^3}\right)^3 =$

28) $\left(\frac{9y}{3y^2}\right)^3 =$

29) $\left(\frac{6x^3y^4}{2x^4y^2}\right)^3 =$

30) $\left(\frac{27x^4y^4}{54x^3y^5}\right)^2 =$

31) $\left(\frac{9x^8y^4}{3x^5y^2}\right)^2 =$

32) $\left(\frac{35\ ^7y^4}{7x^5y^3}\right)^2 =$

✎ **Evaluate each expression. (Zero and Negative Exponents)**

33) $\left(\frac{1}{8}\right)^{-3} =$

34) $\left(\frac{1}{6}\right)^{-2} =$

35) $\left(\frac{3}{4}\right)^{-2} =$

36) $\left(\frac{4}{9}\right)^{-2} =$

37) $\left(\frac{1}{4}\right)^{-4} =$

38) $\left(\frac{2}{7}\right)^{-3} =$

✎ **Write each expression with positive exponents.**

39) $18x^{-2}y^{-6} =$

40) $35x^{-3}y^{-5} =$

41) $-12y^{-4} =$

42) $-25x^{-6} =$

43) $15a^{-3}b^6 =$

44) $20a^6b^{-5}c^{-3} =$

45) $46x^6y^{-3}z^{-7} =$

46) $\frac{16y}{x^3y^{-3}} =$

47) $\frac{24^{-3}b}{-16^{-3}}$

✎ **Write each number in scientific notation.**

48) $0.00521 =$

49) $0.000067 =$

50) $25,000 =$

51) $36,000,000 =$

✎ **Evaluate.**

52) $\sqrt{6} \times \sqrt{6} =$ ....

53) $\sqrt{49} - \sqrt{4} =$ ....

54) $\sqrt{36} + \sqrt{64} =$ ....

55) $\sqrt{9} \times \sqrt{49} =$ ....

56) $\sqrt{2} \times \sqrt{18} =$ ....

57) $3\sqrt{5} + 2\sqrt{5} =$ ....

CHAPTER 9: ANSWERS

1) $8x^4y^2$

2) $48x^4y^3$

3) $15x^5y^5$

4) $28x^3y^5$

5) $27x^7y^7$

6) $42x^6y^5$

7) $8x^7y^8$

8) $21x^7y^5$

9) $100x^9y^5$

10) $40x^8y^5$

11) $18x^6y^8$

12) $49x^6y^2$

13) $\frac{1}{27}$

14) $\frac{1}{5x}$

15) $3x$

16) $\frac{3}{4y^2}$

17) $\frac{1}{3x^4y}$

18) $3xy^2$

19) $\frac{32}{343}$

20) $\frac{6}{x}$

21) $5x^3$

22) $\frac{6y}{7x^5}$

23) $27x^6y^{18}$

24) $32x^{15}y^{20}$

25) $100x^4y^4$

26) $36x^2y^6$

27) $\frac{512}{x^6}$

28) $\frac{27}{y^3}$

29) $\frac{27y^6}{x^3}$

30) $\frac{x^2}{4y^2}$

31) $9x^6y^4$

32) $25x^4y^2$

33) 512

34) 36

35) $\frac{16}{9}$

36) $\frac{81}{16}$

37) 256

38) $\frac{343}{8}$

39) $\frac{18}{x^2y^6}$

40) $\frac{35}{x^3y^5}$

41) $-\frac{12}{y^4}$

42) $-\frac{25}{x^6}$

43) $\frac{15}{a^3}^{6}$

44) $\frac{20a^6}{b^5c^3}$

45) $\frac{46x^6}{y^3z^7}$

46) $\frac{16y^4}{x^3}$

47) $-\frac{3bc^3}{2a^3}$

48) 5.21×10^{-3}

49) 6.7×10^{-5}

50) 25×10^3

51) 36×10^6

52) 6

53) 5

54) 14

55) 21

56) 6

57) $5\sqrt{5}$

CHAPTER 10:

POLYNOMIALS

Math Topics that you'll learn in this chapter:

▶ Simplifying Polynomials

▶ Adding and Subtracting Polynomials

▶ Multiplying Monomials

▶ Multiplying and Dividing Monomials

▶ Multiplying a Polynomial and a Monomial

▶ Multiplying Binomials

▶ Factoring Trinomials

SIMPLIFYING POLYNOMIALS

☑ To simplify Polynomials, find "like" terms. (they have same variables with same power).

☑ Use "FOIL". (First–Out–In–Last) for binomials:

$$(x + a)(x + b) = x^2 + (b + a)x + ab$$

☑ Add or Subtract "like" terms using order of operation.

Examples:

Example 1. Simplify this expression. $x(2x + 5) + 6x =$

Solution: Use Distributive Property: $x(2x + 5) = 2x^2 + 5x$
Now, combine like terms: $x(2x + 5) + 6x = 2x^2 + 5x + 6x = 2x^2 + 11x$

Example 2. Simplify this expression. $(x + 2)(x + 3) =$

Solution: First, apply the FOIL method: $(a + b)(c + d) = ac + ad + bc + bd$
$(x + 2)(x + 3) = x^2 + 3x + 2x + 6$
Now combine like terms: $x^2 + 3x + 2x + 6 = x^2 + 5x + 6$

Example 3. Simplify this expression. $4x(2x - 3) + 6x^2 - 4x =$

Solution: Use Distributive Property: $4x(2x - 3) = 8x^2 - 12x$
Then: $4x(2x - 3) + 6x^2 - 4x = 8x^2 - 12x + 6x^2 - 4x$
Now combine like terms: $8x^2 + 6x^2 = 14x^2$, and $-12x - 4x = -16x$
The simplified form of the expression: $8x^2 - 12x + 6x^2 - 4x = 14x^2 - 16x$

ADDING AND SUBTRACTING POLYNOMIALS

☑ Adding polynomials is just a matter of combining like terms, with some order of operations considerations thrown in.

☑ Be careful with the minus signs, and don't confuse addition and multiplication!

☑ For subtracting polynomials, sometimes you need to use the Distributive Property: $a(b + c) = ab + ac$, $a(b - c) = ab - ac$

Examples:

Example 1. Simplify the expressions. $(x^3 - 3x^4) - (2x^4 - 5x^3) =$

Solution: First, use Distributive Property:

$-(2x^4 - 5x^3) = -1(2x^4 - 5x^3) = -2x^4 + 5x^3$

$\rightarrow (x^3 - 3x^4) - (2x^4 - 5x^3) = x^3 - 3x^4 - 2x^4 + 5x^3$

Now combine like terms: $x^3 + 5x^3 = 6x^3$ and $-3x^4 - 2x^4 = -5x^4$

Then: $(x^3 - 3x^4) - (2x^4 - 5x^3) = x^3 - 3x^4 - 2x^4 + 5x^3 = 6x^3 - 5x^4$

Write the answer in standard form: $6x^3 - 5x^4 = -5x^4 + 6x^3$

Example 2. Add expressions. $(2x^3 - 4) + (6x^3 - 2x^2) =$

Solution: Remove parentheses:

$(2x^3 - 4) + (6x^3 - 2x^2) = 2x^3 - 4 + 6x^3 - 2x^2$

Now combine like terms: $2x^3 - 4 + 6x^3 - 2x^2 = 8x^3 - 2x^2 - 4$

Example 3. Simplify the expressions. $(8x^2 - 3x^3) - (2x^2 + 5x^3) =$

Solution: First, use Distributive Property:

$-(2x^2 + 5x^3) = -2x^2 - 5x^3 \rightarrow (8x^2 - 3x^3) - (2x^2 + 5x^3) = 8x^2 - 3x^3 - 2x^2 - 5x^3$

Now combine like terms and write in standard form:

$8x^2 - 3x^3 - 2x^2 - 5x^3 = -8x^3 + 6x^2$

Multiplying Monomials

☑ A monomial is a polynomial with just one term: Examples: $2x$ or $7y^2$.

☑ When you multiply monomials, first multiply the coefficients (a number placed before and multiplying the variable) and then multiply the variables using multiplication property of exponents.

$$x^a \times x^b = x^{a+b}$$

Examples:

Example 1. Multiply expressions. $5xy^4z^2 \times 3x^2y^5z^3$

Solution: Find the same variables and use multiplication property of exponents: $x^a \times x^b = x^{a+b}$

$x \times x^2 = x^{1+2} = x^3$, $y^4 \times y^5 = y^{4+5} = y^9$ and $z^2 \times z^3 = z^{2+3} = z^5$

Then, multiply coefficients and variables: $5xy^4z^2 \times 3x^2y^5z^3 = 15x^3y^9z^5$

Example 2. Multiply expressions. $-2a^5b^4 \times 8a^3b^4 =$

Solution: Use the multiplication property of exponents: $x^a \times x^b = x^{a+b}$

$a^5 \times a^3 = a^{5+3} = a^8$ and $b^4 \times b^4 = b^{4+4} = b^8$

Then: $-2a^5b^4 \times 8a^3b^4 = -16a^8b^8$

Example 3. Multiply. $7xy^3z^5 \times 4x^2y^4z^3$

Solution: Use the multiplication property of exponents: $x^a \times x^b = x^{a+b}$

$x \times x^2 = x^{1+2} = x^3$, $y^3 \times y^4 = y^{3+4} = y^7$ and $z^5 \times z^3 = z^{5+3} = z^8$

Then: $7xy^3z^5 \times 4x^2y^5z^3 = 28x^3y^7z^8$

Example 4. Simplify. $(5a^6b^3)(-9a^7b^2) =$

Solution: Use the multiplication property of exponents: $x^a \times x^b = x^{a+b}$

$a^6 \times a^7 = a^{6+7} = a^{13}$ and $b^3 \times b^2 = b^{3+2} = b^5$

Then: $(5a^6b^3) \times (-9a^6b^2) = -45a^{13}b^5$

MULTIPLYING AND DIVIDING MONOMIALS

☑ When you divide or multiply two monomials, you need to divide or multiply their coefficients and then divide or multiply their variables.

☑ In case of exponents with the same base, for Division, subtract their powers, for Multiplication, add their powers.

☑ Exponent's Multiplication and Division rules:

$$x^a \times x^b = x^{a+b}, \qquad \frac{x^a}{x^b} = x^{a-b}$$

Examples:

Example 1. Multiply expressions. $(-5x^8)(4x^6) =$

Solution: Use multiplication property of exponents:
$x^a \times x^b = x^{a+b} \rightarrow x^8 \times x^6 = x^{14}$
Then: $(-5x^5)(4x^4) = -20x^{14}$

Example 2. Divide expressions. $\frac{14x^5y^4}{2xy^3} =$

Solution: Use division property of exponents:
$\frac{x^a}{x^b} = x^{a-b} \rightarrow \frac{x^5}{x} = x^{5-1} = x^4$ and $\frac{y^4}{y^3} = y$
Then: $\frac{14x^5y^4}{2xy^3} = 7x^4y$

Example 3. Divide expressions. $\frac{56a^8b^3}{8ab^3}$

Solution: Use division property of exponents:
$\frac{x^a}{x^b} = x^{a-b} \rightarrow \frac{a^8}{a} = a^{8-1} = a^7$ and $\frac{b^3}{b^3} = 1$
Then: $\frac{56a^8b^3}{8ab^3} = 7a^7$

MULTIPLYING A POLYNOMIAL AND A MONOMIAL

☑ When multiplying monomials, use the product rule for exponents.

$$x^a \times x^b = x^{a+b}$$

☑ When multiplying a monomial by a polynomial, use the distributive property.

$$a \times (b + c) = a \times b + a \times c = ab + ac$$
$$a \times (b - c) = a \times b - a \times c = ab - ac$$

Examples:

Example 1. Multiply expressions. $5x(3x - 2)$

Solution: Use Distributive Property:
$5x(3x - 2) = 5x \times 3x - 5x \times (-2) = 15x^2 - 10x$

Example 2. Multiply expressions. $x(2x^2 + 3y^2)$

Solution: Use Distributive Property:
$x(2x^2 + 3y^2) = x \times 2x^2 + x \times 3y^2 = 2x^3 + 3xy^2$

Example 3. Multiply. $-4x(-5x^2 + 3x - 6)$

Solution: Use Distributive Property:
$-4x(-5x^2 + 3x - 6) = (-4x)(-5x^2) + (-4x) \times (3x) + (-4x) \times (-6) =$
Now simplify:
$(-4x)(-5x^2) + (-4x) \times (3x) + (-4x) \times (-6) = 20x^3 - 12x^2 + 24x$

MULTIPLYING BINOMIALS

☑ A binomial is a polynomial that is the sum or the difference of two terms, each of which is a monomial.

☑ To multiply two binomials, use the "FOIL" method. (First–Out–In–Last)

$$(x + a)(x + b) = x \times x + x \times b + a \times x + a \times b = x^2 + bx + ax + ab$$

Examples:

Example 1. Multiply Binomials. $(x + 2)(x - 4) =$

Solution: Use "FOIL". (First–Out–In–Last):
$(x + 2)(x - 4) = x^2 - 4x + 2x - 8$
Then combine like terms: $x^2 - 4x + 2x - 8 = x^2 - 2x - 8$

Example 2. Multiply. $(x - 5)(x - 2) =$

Solution: Use "FOIL". (First–Out–In–Last):
$(x - 5)(x - 2) = x^2 - 2x - 5x + 10$
Then simplify: $x^2 - 2x - 5x + 10 = x^2 - 7x + 10$

Example 3. Multiply. $(x - 3)(x + 6) =$

Solution: Use "FOIL". (First–Out–In–Last):
$(x - 3)(x + 6) = x^2 + 6x - 3x - 18$
Then simplify: $x^2 + 6x - 3x - 18 = x^2 + 3x - 18$

Example 4. Multiply Binomials. $(x + 8)(x + 4) =$

Solution: Use "FOIL". (First–Out–In–Last):
$(x + 8)(x + 4) = x^2 + 4x + 8x + 32$
Then combine like terms: $x^2 + 4x + 8x + 32 = x^2 + 12x + 32$

FACTORING TRINOMIALS

To factor trinomials, you can use following methods:

☑ "FOIL": $(x + a)(x + b) = x^2 + (b + a)x + ab$

☑ "Difference of Squares":

$$a^2 - b^2 = (a + b)(a - b)$$
$$a^2 + 2ab + b^2 = (a + b)(a + b)$$
$$a^2 - 2ab + b^2 = (a - b)(a - b)$$

☑ "Reverse FOIL": $x^2 + (b + a)x + ab = (x + a)(x + b)$

Examples:

Example 1. Factor this trinomial. $x^2 - 2x - 8$

Solution: Break the expression into groups. You need to find two numbers that their product is -8 and their sum is -2. (remember "Reverse FOIL": $x^2 + (b + a)x + ab = (x + a)(x + b)$). Those two numbers are 2 and -4. Then: $x^2 - 2x - 8 = (x^2 + 2x) + (-4x - 8)$
Now factor out x from $x^2 + 2x : x(x + 2)$, and factor out -4 from
$-4x - 8: -4(x + 2)$; Then: $(x^2 + 2x) + (-4x - 8) = x(x + 2) - 4(x + 2)$
Now factor out like term: $(x + 2)$. Then: $(x + 2)(x - 4)$

Example 2. Factor this trinomial. $X^2 - 2x - 24$

Solution: Break the expression into groups: $(x^2 + 4x) + (-6x - 24)$
Now factor out x from $x^2 + 4x : x(x + 4)$, and factor out -6 from
$-6x - 24: -6(x + 4)$; Then: $(x + 4) - 6(x + 4)$, now factor out like term:
$(x = 4) \rightarrow x(x + 4) - 6(x + 4) = (x + 4)(x - 6)$

CHAPTER 10: PRACTICES

✍ Simplify each polynomial.

1) $2(5x + 7) =$

2) $6(3x - 9) =$

3) $x(6x + 3) + 4x =$

4) $2x(x + 8) + 6x =$

5) $8x(2x + 1) - 6x =$

6) $5x(4x - 2) + 2x^2 - 1 =$

7) $4x^2 - 6 - 8x(2x + 7) =$

8) $7x^2 + 9 - 3x(x + 4) =$

✍ Add or subtract polynomials.

9) $(5x^2 + 4) + (3x^2 - 6) =$

10) $(2x^2 - 7x) - (4x^2 + 3x) =$

11) $(8x^3 - 5x^2) + (2x^3 - 6x^2) =$

12) $(3x^3 - 6x) - (7x^3 - 2x) =$

13) $(15x^3 + 3x^2) + (12x^2 - 9) =$

14) $(5x^3 - 8) - (2x^3 - 9x^2) =$

15) $(6x^3 + 2x) + (3x^3 - 2x) =$

16) $(3x^3 - 7x) - (4x^3 + 6x) =$

✍ Find the products. (Multiplying Monomials)

17) $6x^2 \times 4x^3 =$

18) $5x^4 \times 6x^3 =$

19) $-5a^4b \times 4ab^3 =$

20) $(-6x^3yz) \times (-5xy^2z^4) =$

21) $-a^5bc \times a^2b^4 =$

22) $7u^3t^2 \times (-8ut) =$

23) $10x^2z \times 4xy^3 =$

24) $12x^3z \times 2xy^5 =$

25) $-4a^3bc \times a^4b^3 =$

26) $8x^6y^2 \times (-10xy) =$

✎ **Simplify each expression. (Multiplying and Dividing Monomials)**

27) $(6x^2y^3)(9x^4y^2) =$

28) $(3x^3y^2)(7x^4y^3) =$

29) $(12x^8y^5)(4x^5y^7) =$

30) $(10a^3b^2)(5a^3b^8) =$

31) $\dfrac{32\ ^4y^2}{8x^3y} =$

32) $\dfrac{48\ ^5y^6}{6x^2y} =$

33) $\dfrac{72\ ^{15}y^{10}}{9x^8y^6} =$

34) $\dfrac{200x^8y^{12}}{5x^4y^8} =$

✎ **Find each product. (Multiplying a Polynomial and a Monomial)**

35) $2x(4x - y) =$

36) $6x(2x + 5y) =$

37) $6x(x - 9y) =$

38) $x(4x^2 + 3x - 8) =$

39) $6x(-2x^2 + 6x + 3) =$

40) $9x(3x^2 - 6x - 10) =$

✎ **Find each product. (Multiplying Binomials)**

41) $(x - 4)(x + 4) =$

42) $(x - 6)(x - 5) =$

43) $(x + 8)(x + 2) =$

44) $(x - 8)(x + 9) =$

45) $(x + 4)(x - 6) =$

46) $(x - 12)(x + 4) =$

✎ **Factor each trinomial.**

47) $x^2 + x - 12 =$

48) $x^2 + 3x - 10 =$

49) $x^2 - 10x - 24 =$

50) $x^2 + 19x + 48 =$

51) $2x^2 - 14x + 24 =$

52) $3x^2 + 3x - 18 =$

CHAPTER 10: ANSWERS

1) $10x + 14$

2) $18x - 54$

3) $6x^2 + 7x$

4) $2x^2 + 22x$

5) $16x^2 + 2x$

6) $22x^2 - 10x - 1$

7) $-12x^2 - 56x - 6$

8) $4x^2 - 12x + 9$

9) $8x^2 - 2$

10) $-2x^2 - 10x$

11) $10x^3 - 11x^2$

12) $-4x^3 - 4x$

13) $15x^3 + 15x^2 - 9$

14) $3x^3 + 9x^2 - 8$

15) $9x^3$

16) $-x^3 - 13x$

17) $24x^5$

18) $30x^7$

19) $-20a^5b^4$

20) $30x^4y^3z^5$

21) $-a^7b^5c$

22) $-56u^4t^3$

23) $40x^3y^3z$

24) $24x^4y^5z$

25) $-4a^7b^4c$

26) $-80x^7y^3$

27) $54x^6y^5$

28) $21x^7y^5$

29) $48x^{13}y^{12}$

30) $50a^6b^{10}$

31) $4xy$

32) $8x^3y^5$

33) $8x^7y^4$

34) $40x^4y^4$

35) $8x^2 - 2xy$

36) $12x^2 + 30xy$

37) $6x^2 - 54xy$

38) $4x^3 + 3x^2 - 8x$

39) $-12x^3 + 36x^2 + 18x$

40) $27x^3 - 54x^2 - 90x$

41) $x^2 - 16$

42) $x^2 - 11x + 30$

43) $x^2 + 10x + 16$

44) $x^2 + x - 72$

45) $x^2 - 2x - 24$

46) $x^2 - 8x - 48$

47) $(x + 4)(x - 3)$

48) $(x + 5)(x - 2)$

49) $(x - 12)(x + 2)$

50) $(x + 16)(x + 3)$

51) $(2x - 8)(x - 3)$

52) $(3x - 6)(x + 3)$

CHAPTER 11:

GEOMETRY AND SOLID FIGURES

Math Topics that you'll learn in this chapter:

► The Pythagorean Theorem

► Triangles

► Polygons

► Circles

► Trapezoids

► Cubes

► Rectangle Prisms

► Cylinder

THE PYTHAGOREAN THEOREM

☑ You can use the Pythagorean Theorem to find a missing side in a right triangle.

☑ In any right triangle: $a^2 + b^2 = c^2$

Examples:

Example 1. Right triangle ABC (not shown) has two legs of lengths 6 cm (AB) and 8 cm (AC). What is the length of the hypotenuse of the triangle (side BC)?

Solution: Use Pythagorean Theorem: $a^2 + b^2 = c^2$, $a = 6$, and $b = 8$

Then: $a^2 + b^2 = c^2 \rightarrow 6^2 + 8^2 = c^2 \rightarrow 36 + 64 = c^2 \rightarrow 100 = c^2 \rightarrow c = \sqrt{100} = 10$

The length of the hypotenuse is 10 cm.

Example 2. Find the hypotenuse of this triangle.

Solution: Use Pythagorean Theorem: $a^2 + b^2 = c^2$

Then: $a^2 + b^2 = c^2 \rightarrow 12^2 + 5^2 = c^2 \rightarrow 144 + 25 = c^2$

$c^2 = 169 \rightarrow c = \sqrt{169} = 13$

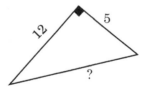

Example 3. Find the length of the missing side in this triangle.

Solution: Use Pythagorean Theorem: $a^2 + b^2 = c^2$

Then: $a^2 + b^2 = c^2 \rightarrow 3^2 + b^2 = 5^2 \rightarrow 9 + b^2 = 25 \rightarrow$

$b^2 = 25 - 9 \rightarrow b^2 = 16 \rightarrow b = \sqrt{16} = 4$

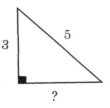

TRIANGLES

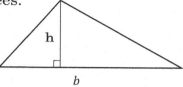

☑ In any triangle, the sum of all angles is 180 degrees.

☑ Area of a triangle $= \frac{1}{2}(base \times height)$

Examples:

What is the area of the following triangles?

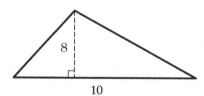

Example 1.

Solution: Use the area formula:
Area $= \frac{1}{2}(base \times height)$
$base = 10$ and $height = 8$
Area $= \frac{1}{2}(10 \times 8) = \frac{1}{2}(80) = 40$

Example 2.

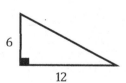

Solution: Use the area formula:
Area $= \frac{1}{2}(base \times height)$
$base = 12$ and $height = 6$; Area $= \frac{1}{2}(12 \times 6) = \frac{72}{2} = 36$

Example 3. What is the missing angle in this triangle?

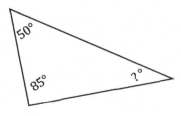

Solution:

In any triangle, the sum of all angles is 180 degrees. Let x be the missing angle.
Then: $50 + 85 + x = 180$;
$\rightarrow 135 + x = 180 \rightarrow x = 180 - 135 = 45$
The missing angle is 45 degrees.

POLYGONS

⊘ The perimeter of a square $= 4 \times side = 4s$ s

⊘ The perimeter of a rectangle$= 2(width + length)$ 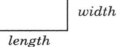 $width$

$length$

⊘ The perimeter of trapezoid$= a + b + c + d$

⊘ The perimeter of a regular hexagon $= 6a$

⊘ The perimeter of a parallelogram $= 2(l + w)$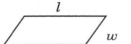

Examples:

Example 1. Find the perimeter of following regular hexagon.

Solution: Since the hexagon is regular, all sides are equal.
Then: The perimeter of The hexagon $= 6 \times (one\ side)$
The perimeter of The hexagon $= 6 \times (one\ side) = 6 \times 4 = 24\ m$

Example 2. Find the perimeter of following trapezoid.

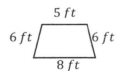

Solution: The perimeter of a trapezoid $= a + b + c + d$
The perimeter of the trapezoid $= 5 + 6 + 6 + 8 = 25\ ft$

CIRCLES

☑ In a circle, variable r is usually used for the radius and d for diameter.

☑ *Area of a circle* $= \pi r^2$ (π is about 3.14)

☑ *Circumference of a circle* $= 2\pi r$

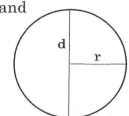

Examples:

Example 1. Find the area of this circle.

Solution:
Use area formula: $Area = \pi r^2$
$r = 8\ in \rightarrow Area = \pi(8)^2 = 64\pi$, $\pi = 3.14$
Then: $Area = 64 \times 3.14 = 200.96\ in^2$

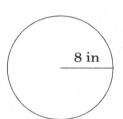

Example 2. Find the Circumference of this circle.

Solution:
Use Circumference formula: $Circumference = 2\pi r$
$r = 5\ cm \rightarrow Circumference = 2\pi(5) = 10\pi$
$\pi = 3.14$ Then: $Circumference = 10 \times 3.14 = 31.4\ cm$

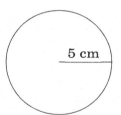

Example 3. Find the area of the circle.

Solution:
Use area formula: $Area = \pi r^2$,
$r = 5\ in$ then: $Area = \pi(5)^2 = 25\pi$, $\pi = 3.14$
Then: $Area = 25 \times 3.14 = 78.5$

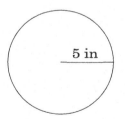

TRAPEZOIDS

☑ A quadrilateral with at least one pair of parallel sides is a trapezoid.

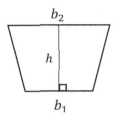

☑ Area of a trapezoid $= \frac{1}{2}h(b_1 + b_2)$

Examples:

Example 1. Calculate the area of this trapezoid.

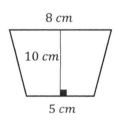

Solution:

Use area formula: $A = \frac{1}{2}h(b_1 + b_2)$

$b_1 = 5 \, cm$, $b_2 = 8 \, cm$ and $h = 10 \, cm$

Then: $A = \frac{1}{2}(10)(8 + 5) = 5(13) = 65 \, cm^2$

Example 2. Calculate the area of this trapezoid.

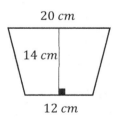

Solution:

Use area formula: $A = \frac{1}{2}h(b_1 + b_2)$

$b_1 = 12 \, cm$, $b_2 = 20 \, cm$ and $h = 14 \, cm$

Then: $A = \frac{1}{2}(14)(12 + 20) = 7(32) = 224 \, cm^2$

CUBES

☑ A cube is a three-dimensional solid object bounded by six square sides.

☑ Volume is the measure of the amount of space inside of a solid figure, like a cube, ball, cylinder or pyramid.

☑ The volume of a cube = $(one\ side)^3$

☑ The surface area of a cube = $6 \times (one\ side)^2$

Examples:

Example 1. Find the volume and surface area of this cube.

Solution: Use volume formula: $volume = (one\ side)^3$
Then: $volume = (one\ side)^3 = (2)^3 = 8\ cm^3$
Use surface area formula:
$surface\ area\ of\ cube$: $6(one\ side)^2 = 6(2)^2 = 6(4) = 24\ cm^2$

2 cm

Example 2. Find the volume and surface area of this cube.

Solution: Use volume formula: $volume = (one\ side)^3$
Then: $volume = (one\ side)^3 = (5)^3 = 125\ cm^3$
Use surface area formula:
$surface\ area\ of\ cube$: $6(one\ side)^2 = 6(5)^2 = 6(25) = 150\ cm^2$

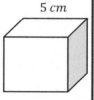
5 cm

Example 3. Find the volume and surface area of this cube.

Solution: Use volume formula: $volume = (one\ side)^3$
Then: $volume = (one\ side)^3 = (7)^3 = 343\ m^3$
Use surface area formula:
$surface\ area\ of\ cube$: $6(one\ side)^2 = 6(7)^2 = 6(49) = 294\ m^2$

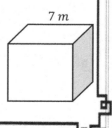
7 m

RECTANGULAR PRISMS

☑ A rectangular prism is a solid 3-dimensional object with six rectangular faces.

☑ The volume of a Rectangular prism = $Length \times Width \times Height$

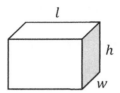

$Volume = l \times w \times h$
$Surface\ area = 2 \times (wh + lw + lh)$

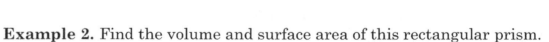

Examples:

Example 1. Find the volume and surface area of this rectangular prism.

Solution: Use volume formula: $Volume = l \times w \times h$
Then: $Volume = 8 \times 6 \times 10 = 480\ m^3$
Use surface area formula: $Surface\ area = 2 \times (wh + lw + lh)$
Then: $Surface\ area = 2 \times \big((6 \times 10) + (8 \times 6) + (8 \times 10)\big)$
$$= 2 \times (60 + 48 + 80) = 2 \times (188) = 376\ m^2$$

Example 2. Find the volume and surface area of this rectangular prism.

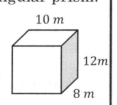

Solution: Use volume formula: $Volume = l \times w \times h$
Then: $Volume = 10 \times 8 \times 12 = 960\ m^3$
Use surface area formula: $Surface\ area = 2 \times (wh + lw + lh)$
Then: $Surface\ area = 2 \times \big((8 \times 12) + (10 \times 8) + (10 \times 12)\big)$
$$= 2 \times (96 + 80 + 120) = 2 \times (296) = 592\ m^2$$

CYLINDER

☑ A cylinder is a solid geometric figure with straight parallel sides and a circular or oval cross-section.

☑ *Volume of a Cylinder* $= \pi(radius)^2 \times height$, $\pi \approx 3.14$

☑ *Surface area of a cylinder* $= 2\pi r^2 + 2\pi rh$

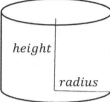

Examples:

Example 1. Find the volume and Surface area of the follow Cylinder.

Solution: Use volume formula:

$Volume = \pi(radius)^2 \times height$

Then: $Volume = \pi(3)^2 \times 8 = 9\pi \times 8 = 72\pi$

$\pi = 3.14$ then: $Volume = 72\pi = 72 \times 3.14 = 226.08 \ cm^3$

Use surface area formula: $Surface \ area = 2\pi r^2 + 2\pi rh$

Then: $2\pi(3)^2 + 2\pi(3)(8) = 2\pi(9) + 2\pi(24) = 18\pi + 48\pi = 66\pi$

$\pi = 3.14$ Then: $Surface \ area = 66 \times 3.14 = 207.24 \ cm^2$

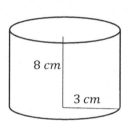

Example 2. Find the volume and Surface area of the follow Cylinder.

Solution: Use volume formula:

$Volume = \pi(radius)^2 \times height$

Then: $Volume = \pi(2)^2 \times 6 = \pi 4 \times 6 = 24\pi$

$\pi = 3.14$ then: $Volume = 24\pi = 75.36 \ cm^3$

Use surface area formula: $Surface \ area = 2\pi r^2 + 2\pi rh$

Then: $= 2\pi(2)^2 + 2\pi(2)(6) = 2\pi(4) + 2\pi(12) = 8\pi + 24\pi = 32\pi$

$\pi = 3.14$ then: $Surface \ area = 32 \times 3.14 = 100.48 \ cm^2$

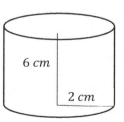

CHAPTER 11: PRACTICES

✍ Find the missing side?

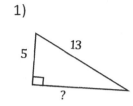
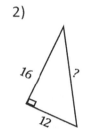
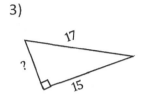
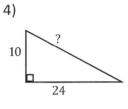

1) 5, 13, ?

2) 16, ?, 12

3) 17, ?, 15

4) 10, ?, 24

✍ Find the measure of the unknown angle in each triangle.

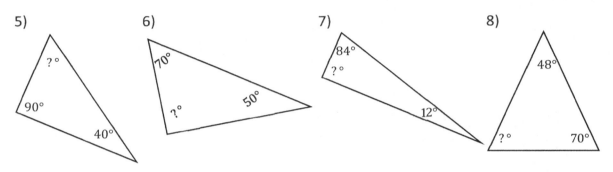

5) ?°, 90°, 40°

6) 70°, ?°, 50°

7) 84°, ?°, 12°

8) 48°, ?°, 70°

✍ Find the area of each triangle.

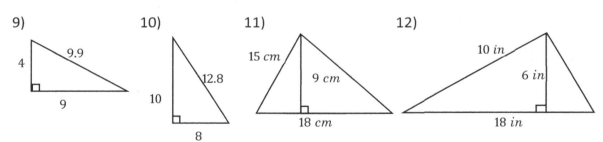

9) 4, 9.9, 9

10) 10, 12.8, 8

11) 15 cm, 9 cm, 18 cm

12) 10 in, 6 in, 18 in

✍ Find the perimeter or circumference of each shape.

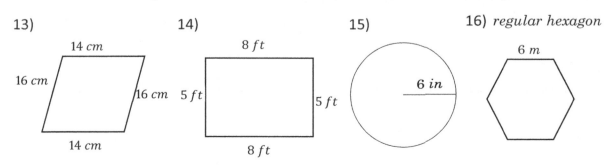

13) 14 cm, 16 cm, 16 cm, 14 cm

14) 8 ft, 5 ft, 5 ft, 8 ft

15) 6 in

16) regular hexagon — 6 m

✍ **Find the area of each trapezoid.**

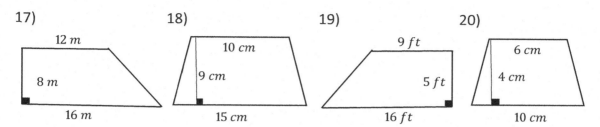

17)
12 m
8 m
16 m

18)
10 cm
9 cm
15 cm

19)
9 ft
5 ft
16 ft

20)
6 cm
4 cm
10 cm

✍ **Find the volume of each cube.**

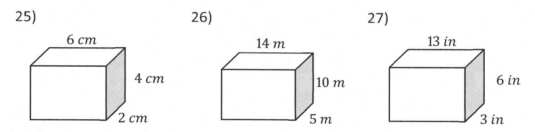

21)
6 cm

22)
9 ft

23)
3 in

24)
8 miles

✍ **Find the volume of each Rectangular Prism.**

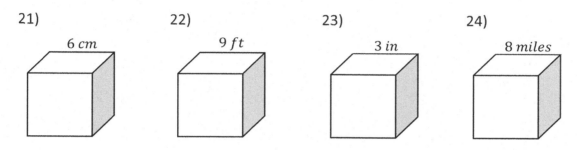

25)
6 cm
4 cm
2 cm

26)
14 m
10 m
5 m

27)
13 in
6 in
3 in

✍ **Find the volume of each Cylinder. Round your answer to the nearest tenth.** ($\pi = 3.14$)

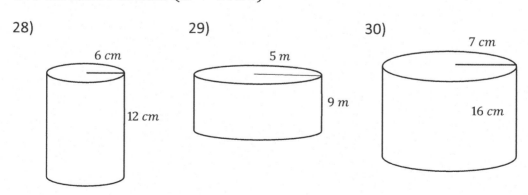

28)
6 cm
12 cm

29)
5 m
9 m

30)
7 cm
16 cm

CHAPTER 11: ANSWERS

1) 12

2) 20

3) 8

4) 26

5) 50

6) 60

7) 84

8) 62

9) 18

10) 40

11) $81\ cm^2$

12) $54 in^2$

13) $60\ cm$

14) $26\ ft$

15) $12\ \pi \approx 37.68\ in$

16) $36\ m$

17) $112\ m^2$

18) $112.5\ cm^2$

19) $62.5\ ft^2$

20) $32\ cm^2$

21) $216\ cm^3$

22) $729\ ft^3$

23) $27\ in^3$

24) $512\ mi^3$

25) $48\ cm^3$

26) $700\ m^3$

27) $234\ in^3$

28) $1,356.5\ cm^3$

29) $706.5\ m^3$

30) $2,461.8\ cm^3$

CHAPTER 12:

STATISTICS

Math Topics that you'll learn in this chapter:

▶ Mean, Median, Mode, and Range of the Given Data

▶ Pie Graph

▶ Probability Problems

▶ Permutations and Combinations

MEAN, MEDIAN, MODE, AND RANGE OF THE GIVEN DATA

☑ Mean: $\dfrac{sum\ of\ the\ data}{total\ number\ of\ data\ entires}$

☑ Mode: the value in the list that appears most often

☑ Median: is the middle number of a group of numbers arranged in order by size.

☑ Range: the difference of the largest value and smallest value in the list

Examples:

Example 1. What is the mode of these numbers? $4, 5, 7, 5, 7, 4, 0, 4$

Solution: Mode: the value in the list that appears most often.
Therefore, the mode is number 4. There are three number 4 in the data.

Example 2. What is the median of these numbers? $5, \quad , \quad , 9, \quad , \quad , 6$

Solution: Write the numbers in order: $5, 6, 9, \quad , \quad , $
The median is the number in the middle. Therefore, the median is   .

Example 3. What is the mean of these numbers? $8, 2, 8, 5, 3, 2, 4, 8$

Solution: Mean: $\dfrac{the}{\quad} = \dfrac{8+2+8+5+3+2+4+8}{8} = 5$

Example 4. What is the range in this list? $4, 9, \quad , 8, \quad , \quad , 5$

Solution: Range is the difference of the largest value and smallest value in the list. The largest value is 18 and the smallest value is 4.
Then: $\quad - 4 = $

PIE GRAPH

☑ A Pie Chart is a circle chart divided into sectors, each sector represents the relative size of each value.

☑ Pie charts represent a snapshot of how a group is broken down into smaller pieces.

Example:

A library has 820 books that include Mathematics, Physics, Chemistry, English and History. Use the following graph to answer the questions.

Example 1. What is the number of Mathematics books?

Solution: Number of total books = 820

Percent of Mathematics books = 30% = 0.30

Then, the number of Mathematics books: 0.30 × 820 = 246

Example 2. What is the number of History books?

Solution: Number of total books = 820

Percent of History books = 10% = 0.10

Then: 0.10 × 820 = 82

Example 3. What is the number of Chemistry books?

Solution: Number of total books = 820

Percent of Chemistry books = 20% = 0.20

Then: 0.20 × 820 = 164

Probability Problems

☑ Probability is the likelihood of something happening in the future. It is expressed as a number between zero (can never happen) to 1 (will always happen).

☑ Probability can be expressed as a fraction, a decimal, or a percent.

☑ Probability formula: $Probability = \dfrac{number\ of\ desired\ outcomes}{number\ of\ total\ outcomes}$

Examples:

Example 1. Anita's trick–or–treat bag contains 12 pieces of chocolate, 18 suckers, 18 pieces of gum, 24 pieces of licorice. If she randomly pulls a piece of candy from her bag, what is the probability of her pulling out a piece of sucker?

Solution: Probability $= \dfrac{number\ of\ desired\ outcomes}{number\ of\ total\ outcomes}$

Probability of pulling out a piece of sucker $= \dfrac{18}{12 + 18 + 18 + 24} = \dfrac{18}{72} = \dfrac{1}{4}$

Example 2. A bag contains 20 balls: four green, five black, eight blue, a brown, a red and one white. If 19 balls are removed from the bag at random, what is the probability that a brown ball has been removed?

Solution: If 19 balls are removed from the bag at random, there will be one ball in the bag. The probability of choosing a brown ball is 1 out of 20. Therefore, the probability of not choosing a brown ball is 19 out of 20 and the probability of having not a brown ball after removing 19 balls is the same.

PERMUTATIONS AND COMBINATIONS

☑ Factorials are products, indicated by an exclamation mark. For example, $4! = 4 \times 3 \times 2 \times 1$ (Remember that $0!$ is defined to be equal to 1.)

☑ Permutations: The number of ways to choose a sample of k elements from a set of n distinct objects where order does matter, and replacements are not allowed. For a permutation problem, use this formula:

$$_nPk = \frac{n!}{(n-k)!}$$

☑ Combination: The number of ways to choose a sample of r elements from a set of n distinct objects where order does not matter, and replacements are not allowed. For a combination problem, use this formula:

$$_nCr = \frac{n!}{r!\,(n-r)!}$$

Examples:

Example 1. How many ways can the first and second place be awarded to 8 people?

Solution: Since the order matters, (the first and second place are different!) we need to use permutation formula where n is 10 and k is 2. Then: $\frac{n!}{(n-k)!} = \frac{8!}{(8-2)!} = \frac{8!}{6!} = \frac{8\times7\times6!}{6!}$, remove 6! from both sides of the fraction. Then: $\frac{8\times7\times6!}{6!} = 8 \times 7 = 56$

Example 2. How many ways can we pick a team of 2 people from a group of 6?

Solution: Since the order doesn't matter, we need to use a combination formula where n is 8 and r is 3. Then: $\frac{n!}{r!\,(n-r)!} = \frac{6!}{2!\,(6-2)!} = \frac{6!}{2!\,(4)!} = \frac{6\times5\times4!}{2!\,(4)!} = \frac{6\times5}{2\times1} = \frac{30}{2} = 15$

CHAPTER 12: PRACTICES

✎ Find the values of the Given Data.

1) 7, 10, 4, 2, 7

Mode: _____   Range: _____

Mean: _____   Median: _____

2) 4, 8, 2, 9, 8, 5

Mode: _____   Range: _____

Mean: _____   Median: _____

3) 12, 2, 6, 10, 6, 15

Mode: _____   Range: _____

Mean: _____   Median: _____

4) 12, 5, 1, 10, 2, 11, 1

Mode: _____   Range: _____

Mean: _____   Median: _____

✎ The circle graph below shows all Bob's expenses for last month. Bob spent $896 on his Rent last month.

5) How much did Bob's total expenses last month? _____

6) How much did Bob spend for foods last month? _____

7) How much did Bob spend for his bills last month? _____

8) How much did Bob spend on his car last month? _____

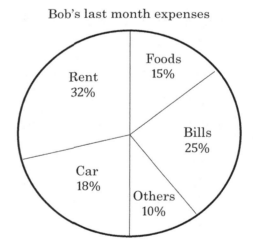

Bob's last month expenses

✍ Solve.

9) Bag A contains 6 red marbles and 9 green marbles. Bag B contains 4 black marbles and 7 orange marbles. What is the probability of selecting a green marble at random from bag A? What is the probability of selecting a black marble at random from Bag B?

_____   _____

✍ Solve.

10) Susan is baking cookies. She uses sugar, flour, butter, and eggs. How many different orders of ingredients can she try? _____

11) Jason is planning for his vacation. He wants to go to museum, go to the beach, and play volleyball. How many different ways of ordering are there for him? _____

12) In how many ways can a team of 8 basketball players choose a captain and co-captain? _____

13) How many ways can you give 6 balls to your 8 friends? _____

14) A professor is going to arrange her 6 students in a straight line. In how many ways can she do this? _____

15) In how many ways can a teacher chooses 5 out of 13 students?

_____

CHAPTER 12: ANSWERS

1) Mode: 7, Range: 8, Mean: 6, Median: 7

2) Mode: 8, Range:7, Mean: 6, Median: 6.5

3) Mode: 6, Range: 13, Mean: 8.5, Median: 8

4) Mode: 1, Range: 11, Mean: 6, Median: 6

5) $2,800

6) $420

7) $700

8) $504

9) $\frac{3}{5}, \frac{4}{11}$

10) 24

11) 6

12) 56 (it's a permutation problem)

13) 28 (it's a combination problem)

14) 720

15) 1,287 (it's a combination problem)

CHAPTER 13:

FUNCTIONS OPERATIONS

Math Topics that you'll learn in this chapter:

☑ Function Notation

☑ Adding and Subtracting Functions

☑ Multiplying and Dividing Functions

☑ Composition of Functions

FUNCTION NOTATION AND EVALUATION

☑ Functions are mathematical operations that assign unique outputs to given inputs.

☑ Function notation is the way a function is written. It is meant to be a precise way of giving information about the function without a rather lengthy written explanation.

☑ The most popular function notation is $f(x)$ which is read "f of x". Any letter can name a function. for example: $g(x)$, $h(x)$, etc.

☑ To evaluate a function, plug in the input (the given value or expression) for the function's variable (place holder, x).

Examples:

Example 1. Evaluate: $h(n) = 2n - 2$, find $h(2)$

Solution: Substitute n with 2:
Then: $h(n) = 2n - 2 \rightarrow h(2) = 2(2) - 2 \rightarrow h(2) = 4 - 2 = 2$

Example 2. Evaluate: $w(x) = 5x - 1$, find $w(3)$.

Solution: Substitute x with 3:
Then: $w(x) = 5x - 1 \rightarrow w(3) = 5(3) - 1 = 15 - 1 = 14$

Example 3. Evaluate: $f(x) = x^2 - 2$, find $f(2)$.

Solution: Substitute x with 2:
Then: $f(x) = x^2 - 2 \rightarrow f(2) = (2)^2 - 2 = 4 - 2 = 2$

Example 4. Evaluate: $p(x) = 2x^2 - 4$, find $p(3a)$.

Solution: Substitute x with $3a$:
Then: $p(x) = 2x^2 - 4 \rightarrow p(3a) = 2(3a)^2 - 4 \rightarrow p(3a) = 2(9a^2) - 4 = 18a^2 - 4$

ADDING AND SUBTRACTING FUNCTIONS

☑ Just like we can add and subtract numbers and expressions, we can add or subtract two functions and simplify or evaluate them. The result is a new function.

☑ For two functions $f(x)$ and $g(x)$, we can create two new functions:

$(f + g)(x) = f(x) + g(x)$ and $(f - g)(x) = f(x) - g(x)$

Examples:

Example 1. $g(x) = a - 1$, $f(a) = a + 2$, Find: $(g + f)(a)$

Solution: $(g + f)(a) = g(a) + f(a)$
Then: $(g + f)(a) = (a - 1) + (a + 2) = 2a + 1$

Example 2. $f(x) = 2x - 2$, $g(x) = x - 4$, Find: $(f - g)(x)$

Solution: $(f - g)(x) = f(x) - g(x)$
Then: $(f - g)(x) = (2x - 2) - (x - 4) = 2x - 2 - x + 4 = x + 2$

Example 3. $g(x) = x^2 - 4$, $f(x) = 2x + 3$, Find: $(g + f)(x)$

Solution: $(g + f)(x) = g(x) + f(x)$
Then: $(g + f)(x) = (x^2 - 4) + (2x + 3) = x^2 + 2x - 1$

Example 4. $f(x) = 2x^2 + 5$, $g(x) = 3x - 1$, Find: $(f - g)(5)$

Solution: $(f - g)(x) = f(x) - g(x)$
Then: $(f - g)(x) = (2x^2 + 5) - (3x - 1) = 2x^2 + 5 - 3x + 1 = 2x^2 - 3x + 6$
Substitute x with 5: $(g - f)(5) = 2(5)^2 - 3(5) + 6 = 50 - 15 + 6 = 41$

MULTIPLYING AND DIVIDING FUNCTIONS

☑ Just like we can multiply and divide numbers and expressions, we can multiply and divide two functions and simplify or evaluate them.

☑ For two functions $f(x)$ and $g(x)$, we can create two new functions:

$$(f.g)(x) = f(x).g(x) \text{ and } \left(\frac{f}{g}\right)(x) = \frac{f(x)}{g(x)}$$

Examples:

Example 1. $g(x) = x - 2$, $f(x) = x + 3$, Find: $(g.f)(x)$

Solution: $(g.f)(x) = g(x).f(x) = (x - 2)(x + 3) = x^2 + 3x - 2x - 6$

Example 2. $f(x) = x + 4$, $h(x) = x - 6$, Find: $\left(\frac{f}{h}\right)(x)$

Solution: $\left(\frac{f}{h}\right)(x) = \frac{f(x)}{h(x)} = \frac{x+4}{x-6}$

Example 3. $g(x) = x + 5$, $f(x) = x - 2$, Find: $(g.f)(4)$

Solution: $(g.f)(x) = g(x).f(x) = (x + 5)(x - 2) = x^2 - 2x + 5x - 10$
$g(x).f(x) = x^2 + 3x - 10$
Substitute x with 4: $(g.f)(x) = (4)^2 + 3(4) - 10 = 16 + 12 - 10 = 18$

Example 4. $f(x) = 2x + 3$, $h(x) = x + 8$, Find: $\left(\frac{f}{h}\right)(-1)$

Solution: $\left(\frac{f}{h}\right)(x) = \frac{f(x)}{h(x)} = \frac{2x+3}{x+8}$
Substitute x with -1: $\left(\frac{f}{h}\right)(x) = \frac{2x+3}{x+8} = \frac{2(-1)+3}{(-1)+8} = \frac{1}{7}$

COMPOSITION OF FUNCTIONS

☑ "Composition of functions" simply means combining two or more functions in a way where the output from one function becomes the input for the next function.

☑ The notation used for composition is: $(fog)(x) = f(g(x))$ and is read

"f composed with g of x" or "f of g of x".

Examples:

Example 1. *Using* f(x) = x − 5 *and* g(x) = 2x, *find:* (fog)(x)

Solution: $(fog)(x) = f(g(x))$. Then: $(fog)(x) = f(g(x)) = f(2x)$
Now find $f(2x)$ by substituting x with $2x$ in $f(x)$ function.
Then: f(x) = x − 5 ; $(x \rightarrow 2x) \rightarrow f(2x) = (2x) - 5 = 2x - 5$

Example 2. *Using* f(x) = x + 6 *and* g(x) = x − 2, *find:* (g o f)(−1)

Solution: $(f \ o \ g)(x) = f(g(x))$. Then: $(g \ o \ f)(x) = g(f(x)) = g(x + 6)$, now substitute x in g(x) *by* (x + 6).
Then: $g(x + 6) = (x + 6) - 2 = x + 6 - 2 = x + 4$
Substitute x with −1: $(g \ o \ f)(-1) = g(f(x)) = x + 4 = -1 + 4 = 3$

Example 3. *Using* f(x) = 2x − 2 *and* g(x) = 2x, *find:* f(g(5))

Solution: First, find $g(5)$): g(x) = 2x → g(5) = 2(5) = 10
Then: $f(g(5)) = f(10)$. Now, find $f(10)$ by substituting x with 10 in $f(x)$ function. Then: $f(g(5)) = f(10) = 2(10) - 2 = 20 - 2 = 18$

CHAPTER 13: PRACTICES

✍ Evaluate each function.

1) $g(n) = 5n - 2$, find $g(-3)$

_____

2) $h(x) = -4x + 9$, find $h(4)$

_____

3) $k(n) = 10 - 6n$, find $k(2)$

_____

4) $g(x) = 6x - 1$, find $g(-1)$

_____

5) $k(n) = 7n - 3$, find $k(5)$

_____

6) $w(n) = -3n + 10$, find $w(6)$

_____

✍ Perform the indicated operation.

7) $f(x) = x + 3$
   $g(x) = 4x + 1$
   Find $(f - g)(x)$

_____

8) $g(x) = x - 5$
   $f(x) = -x - 6$
   Find $(g - f)(x)$

_____

9) $h(t) = 6t + 2$
   $g(t) = 3t + 4$
   Find $(h + g)(x)$

_____

10) $g(a) = 5a - 3$
    $f(a) = a^2 + 2$
    Find $(g + f)(2)$

_____

11) $g(x) = 2x - 6$
    $h(x) = 3x^2 + 1$
    Find $(g - f)(-4)$

_____

12) $h(x) = x^2 + 6$
    $g(x) = -5x + 1$
    Find $(h + g)(4)$

_____

✎ Perform the indicated operation.

13) $g(x) = x + 1$

$f(x) = x + 2$

Find $(g \cdot f)(x)$

_____

14) $f(x) = 3x$

$h(x) = -x + 4$

Find $(f \cdot h)(x)$

_____

15) $g(a) = a + 5$

$h(a) = 3a - 1$

Find $(g \cdot h)(6)$

_____

16) $f(x) = 2x + 3$

$h(x) = 3x - 1$

Find $\left(\frac{f}{h}\right)(3)$

_____

17) $f(x) = a^2 - 1$

$g(x) = -3 + 4a$

Find $\left(\frac{f}{g}\right)(6)$

_____

18) $g(a) = 3a + 6$

$f(a) = a - 1$

Find $\left(\frac{g}{f}\right)(4)$

_____

✎ Using $f(x) = 3x + 1$ and $g(x) = x - 4$, find:

19) $g\big(f(1)\big) = $ _____

20) $g\big(f(-1)\big) = $ _____

21) $f\big(g(3)\big) = $ _____

22) $f\big(f(6)\big) = $ _____

23) $g\big(f(4)\big) = $ _____

24) $g\big(f(-6)\big) = $ _____

25) $g\big(f(5)\big) = $ ____

26) $g\big(f(-5)\big) = $ ____

27) $f\big(g(-2)\big) = $ ____

CHAPTER 13: ANSWERS

1) -17

2) -7

3) -2

4) -7

5) 32

6) -8

7) $-3x + 2$

8) $-2x - 1$

9) $9t + 6$

10) 13

11) 63

12) 3

13) $x^2 + 3x + 2$

14) $-3x^2 + 12x$

15) 187

16) $\frac{9}{8}$

17) $\frac{35}{21}$

18) 6

19) 0

20) -6

21) -2

22) 58

23) 9

24) -21

25) 12

26) -18

27) -17

ISEE UPPER LEVEL TEST REVIEW

The Independent School Entrance Exam (ISEE) is a standardized test developed by the Educational Records Bureau for its member schools as part of their admission process.

 ❖ There are currently four Levels of the ISEE:
 ❖ Primary Level (entering Grades 2 - 4)
 ❖ Lower Level (entering Grades 5 and 6)
 ❖ Middle Level (entering Grades 7 and 8)
 ❖ Upper Level (entering Grades 9 - 12)

There are five sections on the ISEE Upper Level Test:
 ❖ Verbal Reasoning
 ❖ Quantitative Reasoning
 ❖ Reading Comprehension
 ❖ Mathematics Achievement
 ❖ and a 30-minute essay

ISEE Upper Level tests use a multiple-choice format and contain two Mathematics sections:

Quantitative Reasoning

There are 37 questions in the Quantitative Reasoning section and students have 35 minutes to answer the questions. This section contains word problems and quantitative comparisons. The word problems require either no calculation or simple calculation. The quantitative comparison items present two quantities,

(A) and (B), and the student needs to select one of the following four answer choices:

(A) The quantity in Column A is greater.

(B) The quantity in Column B is greater.

(C) The two quantities are equal.

(D) The relationship cannot be determined from the information given.

Mathematics Achievement

There are 47 questions in the Mathematics Achievement section and students have 40 minutes to answer the questions. Mathematics Achievement measures students' knowledge of Mathematics requiring one or more steps in calculating the answer.

In this book, there are two complete ISEE Upper Level Quantitative Reasoning and Mathematics Achievement Tests. Let your student take these tests to see what score they will be able to receive on a real ISEE Upper Level test.

Good luck!

Time to refine your skill with a practice examination

Take practice ISEE Upper Level Math Tests to simulate the test day experience. After you've finished, score your tests using the answer keys.

Before You Start

❖ You'll need a pencil and a timer to take the test.

❖ For each question, there are four possible answers. Choose which one is best.

❖ After you've finished the test, review the answer key to see where you went wrong.

❖ Use the answer sheet provided to record your answers. (You can cut it out or photocopy it)

❖ You will receive 1 point for every correct answer, and you will lose $\frac{1}{4}$ point for each incorrect answer. There is no penalty for skipping a question.

Calculators are NOT permitted for the ISEE Upper Level Test

Good Luck!

ISEE Upper Level Math Practice Test 1

2020 - 2021

Two Parts

- ▶ **Total number of questions:** 84
- ▶ **Part 1 (Calculator):** 37 questions
- ▶ **Part 2 (Calculator):** 47 questions
- ▶ **Total time for two parts:** 75 Minutes

ISEE Upper Level Practice Tests Answer Sheet

Remove (or photocopy) these answer sheets and use them to complete the practice tests.

ISEE Upper Level Practice Test			
Quantitative Reasoning		Mathematics Achievement	

Quantitative Reasoning

1. Ⓐ Ⓑ Ⓒ Ⓓ   25. Ⓐ Ⓑ Ⓒ Ⓓ
2. Ⓐ Ⓑ Ⓒ Ⓓ   26. Ⓐ Ⓑ Ⓒ Ⓓ
3. Ⓐ Ⓑ Ⓒ Ⓓ   27. Ⓐ Ⓑ Ⓒ Ⓓ
4. Ⓐ Ⓑ Ⓒ Ⓓ   28. Ⓐ Ⓑ Ⓒ Ⓓ
5. Ⓐ Ⓑ Ⓒ Ⓓ   29. Ⓐ Ⓑ Ⓒ Ⓓ
6. Ⓐ Ⓑ Ⓒ Ⓓ   30. Ⓐ Ⓑ Ⓒ Ⓓ
7. Ⓐ Ⓑ Ⓒ Ⓓ   31. Ⓐ Ⓑ Ⓒ Ⓓ
8. Ⓐ Ⓑ Ⓒ Ⓓ   32. Ⓐ Ⓑ Ⓒ Ⓓ
9. Ⓐ Ⓑ Ⓒ Ⓓ   33. Ⓐ Ⓑ Ⓒ Ⓓ
10. Ⓐ Ⓑ Ⓒ Ⓓ   34. Ⓐ Ⓑ Ⓒ Ⓓ
11. Ⓐ Ⓑ Ⓒ Ⓓ   35. Ⓐ Ⓑ Ⓒ Ⓓ
12. Ⓐ Ⓑ Ⓒ Ⓓ   36. Ⓐ Ⓑ Ⓒ Ⓓ
13. Ⓐ Ⓑ Ⓒ Ⓓ   37. Ⓐ Ⓑ Ⓒ Ⓓ
14. Ⓐ Ⓑ Ⓒ Ⓓ
15. Ⓐ Ⓑ Ⓒ Ⓓ
16. Ⓐ Ⓑ Ⓒ Ⓓ
17. Ⓐ Ⓑ Ⓒ Ⓓ
18. Ⓐ Ⓑ Ⓒ Ⓓ
19. Ⓐ Ⓑ Ⓒ Ⓓ
20. Ⓐ Ⓑ Ⓒ Ⓓ
21. Ⓐ Ⓑ Ⓒ Ⓓ
22. Ⓐ Ⓑ Ⓒ Ⓓ
23. Ⓐ Ⓑ Ⓒ Ⓓ
24. Ⓐ Ⓑ Ⓒ Ⓓ

Mathematics Achievement

1. Ⓐ Ⓑ Ⓒ Ⓓ   25. Ⓐ Ⓑ Ⓒ Ⓓ
2. Ⓐ Ⓑ Ⓒ Ⓓ   26. Ⓐ Ⓑ Ⓒ Ⓓ
3. Ⓐ Ⓑ Ⓒ Ⓓ   27. Ⓐ Ⓑ Ⓒ Ⓓ
4. Ⓐ Ⓑ Ⓒ Ⓓ   28. Ⓐ Ⓑ Ⓒ Ⓓ
5. Ⓐ Ⓑ Ⓒ Ⓓ   29. Ⓐ Ⓑ Ⓒ Ⓓ
6. Ⓐ Ⓑ Ⓒ Ⓓ   30. Ⓐ Ⓑ Ⓒ Ⓓ
7. Ⓐ Ⓑ Ⓒ Ⓓ   31. Ⓐ Ⓑ Ⓒ Ⓓ
8. Ⓐ Ⓑ Ⓒ Ⓓ   32. Ⓐ Ⓑ Ⓒ Ⓓ
9. Ⓐ Ⓑ Ⓒ Ⓓ   33. Ⓐ Ⓑ Ⓒ Ⓓ
10. Ⓐ Ⓑ Ⓒ Ⓓ   34. Ⓐ Ⓑ Ⓒ Ⓓ
11. Ⓐ Ⓑ Ⓒ Ⓓ   35. Ⓐ Ⓑ Ⓒ Ⓓ
12. Ⓐ Ⓑ Ⓒ Ⓓ   36. Ⓐ Ⓑ Ⓒ Ⓓ
13. Ⓐ Ⓑ Ⓒ Ⓓ   37. Ⓐ Ⓑ Ⓒ Ⓓ
14. Ⓐ Ⓑ Ⓒ Ⓓ   38. Ⓐ Ⓑ Ⓒ Ⓓ
15. Ⓐ Ⓑ Ⓒ Ⓓ   39. Ⓐ Ⓑ Ⓒ Ⓓ
16. Ⓐ Ⓑ Ⓒ Ⓓ   40. Ⓐ Ⓑ Ⓒ Ⓓ
17. Ⓐ Ⓑ Ⓒ Ⓓ   41. Ⓐ Ⓑ Ⓒ Ⓓ
18. Ⓐ Ⓑ Ⓒ Ⓓ   42. Ⓐ Ⓑ Ⓒ Ⓓ
19. Ⓐ Ⓑ Ⓒ Ⓓ   43. Ⓐ Ⓑ Ⓒ Ⓓ
20. Ⓐ Ⓑ Ⓒ Ⓓ   44. Ⓐ Ⓑ Ⓒ Ⓓ
21. Ⓐ Ⓑ Ⓒ Ⓓ   45. Ⓐ Ⓑ Ⓒ Ⓓ
22. Ⓐ Ⓑ Ⓒ Ⓓ   46. Ⓐ Ⓑ Ⓒ Ⓓ
23. Ⓐ Ⓑ Ⓒ Ⓓ   47. Ⓐ Ⓑ Ⓒ Ⓓ
24. Ⓐ Ⓑ Ⓒ Ⓓ

ISEE Upper Level Math Practice Test 1 Part 1 (Quantitative Reasoning)

Total number of questions: 37

Total time for Part 1: 35 Minutes

You may NOT use a calculator on this part.

1) If $2y + 6 < 30$, then y could be equal to?

A. 15

B. 14

C. 12

D. 8

2) Which of the following is NOT a factor of 90?

A. 9

B. 10

C. 16

D. 30

3) What is the area of a square whose diagonal is 6 meters?

A. $20\ m^2$

B. $18\ m^2$

C. $12\ m^2$

D. $10\ m^2$

4) If Emily left a $13.26 tip on a breakfast that cost $58.56, approximately what percentage was the tip?

A. 24%

B. 22%

C. 20%

D. 18%

5) There are 7 blue marbles, 9 red marbles, and 6 yellow marbles in a box. If Ava randomly selects a marble from the box, what is the probability of selecting a red or yellow marble?

A. $\dfrac{1}{7}$

B. $\dfrac{1}{9}$

C. $\dfrac{15}{22}$

D. $\dfrac{5}{7}$

6) James earns $8.50 per hour and worked 20 hours. Jacob earns $10.00 per hour. How many hours would Jacob need to work to equal James's earnings over 20 hours?

A. 14

B. 17

C. 20

D. 25

7) A phone company charges $5 for the first five minutes of a phone call and 50 cents per minute thereafter. If Sofia makes a phone call that lasts 30 minutes, what will be the total cost of the phone call?

A. 18.00

B. 18.50

C. 20.00

D. 20.50

8) Michelle and Alec can finish a job together in 50 minutes. If Michelle can do the job by herself in 2.5 hours, how many minutes does it take Alec to finish the job?

A. 100

B. 75

C. 50

D. 40

9) If 150% of a number is 75, then what is the 90% of that number?

A. 45

B. 50

C. 60

D. 70

10) In the figure, *MN* is 50 *cm*. How long is *ON*?

A. 35 *cm*

B. 30 *cm*

C. 25 *cm*

D. 20 *cm*

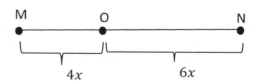

11) On a map, the length of the road from City A to City B is measured to be 18 inches. On this map, $\frac{1}{2}$ inch represents an actual distance of 14 miles. What is the actual distance, in miles, from City A to City B along this road?

A. 504 miles

B. 620 miles

C. 860 miles

D. 1,260 miles

A library has 700 books that include Mathematics, Physics, Chemistry, English and History.

Use following graph to answer questions 12.

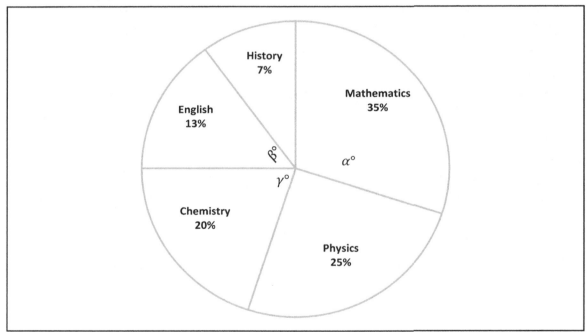

12) What is the product of the number of Mathematics and number of English books?

A. 10,870

B. 18,452

C. 22,295

D. 35,265

13) How many $\frac{1}{6}$ pound paperback books together weigh 60 pounds?

A. 100

B. 200

C. 300

D. 360

14) Emily and Daniel have taken the same number of photos on their school trip. Emily has taken 6 times as many as photos as Claire and Daniel has taken 15 more photos than Claire. How many photos has Claire taken?

A. 2

B. 3

C. 9

D. 11

15) The first four terms in a sequence are shown below. What is the sixth term in the sequence?

1) $\{3, 6, 11, 18, ...\}$

A. 38

B. 40

C. 45

D. 50

16) What is the equation of the line that passes through $(3, -3)$ and has a slope of 2?

A. $y = 2x - 9$

B. $y = 3x - 3$

C. $y = 2x + 3$

D. $y = 3x + 9$

17) Find the solution (x, y) to the following system of equations?
$$-4x - y = 8$$
$$8x + 6y = 20$$

A. $(5, 14)$

B. $(8, 6)$

C. $(17, 11)$

D. $(-\frac{17}{4}, 9)$

18) Sophia purchased a sofa for $530.20. The sofa is regularly priced at $631. What was the percent discount Sophia received on the sofa?

A. 12%

B. 16%

C. 21%

D. 26%

19) Three second of 20 is equal to $\frac{5}{2}$ of what number?

A. 12

B. 20

C. 40

D. 60

20) A supermarket's sales increased by 11 percent in the month of April and decreased by 11 percent in the month of May. What is the percent change in the sales of the supermarket over the two-month period?

A. 2% decrease

B. No change

C. 2% increase

D. 2.2% increase

21) The distance between cities A and B is approximately 2,700 miles. If Alice drives an average of 74 miles per hour, how many hours will it take Alice to drive from city A to city B?

A. *Approximately* 41 *hours*

B. *Approximately* 36 *hours*

C. *Approximately* 28 *hours*

D. *Approximately* 21 *hours*

Quantitative Comparisons

Direction: Questions 22 to 37 are Quantitative Comparisons Questions. Using the information provided in each question, compare the quantity in column A to the quantity in Column B. Choose on your answer sheet grid

A. if the quantity in Column A is greater

B. if the quantity in Column B is greater

C. if the two quantities are equal

D. if the relationship cannot be determined from the information given

22) $x^2 = 8$

Column A	Column B
x	2

23) For all numbers x and y, let the function $x \diamondsuit y$ be defined by $x \diamondsuit y = x^2 - 2xy + y^2$

Column A	Column B
$4 \diamondsuit 5$	$5 \diamondsuit 4$

24)

Column A	Column B
The average of $16, 22, 24, 36, 40$	The average of $28, 33, 38, 42$

25)

Column A	Column B
$\dfrac{1}{x+1}$	$\dfrac{2}{x+2}$

26) $\dfrac{x}{3} = y^2$

Column A	Column B
x	y

27) $x = 1$

Column A

$2x^3 - 3x - 2$

Column B

$3x^2 - 2x - 3$

28)

Column A

$8 + 14(7 - 3)$

Column B

$14 + 8(7 - 3)$

29) $\frac{x}{36} = \frac{3}{4}$

Column A

$\dfrac{9}{x}$

Column B

$\dfrac{1}{9}$

30) Working at constant rates, machine D makes b rolls of steel in 38 minutes and machine E makes b rolls of steel in one hour ($b > 0$)

Column A

The number of rolls of steel made by machine D in 3 hours and 10 minutes.

Column B

The number of rolls of steel made by machine E in 5 hours.

31) $-1 < y < 4$

Column A

$\dfrac{y}{2}$

Column B

$\dfrac{2}{y}$

32) $\frac{a}{b} = \frac{c}{d}$

Column A

$a + b$

Column B

$c + d$

33) The ratio of boys to girls in a class is 5 to 7.

Column A	Column B
Ratio of boys to the entire class	$\frac{1}{3}$

34) There are only 4 blue marbles and 5 green marbles in a jar. Two marbles are pulled out in succession without replacing them in the jar.

Column A	Column B
The probability that both marbles are blue.	The probability that the first marbles is green, but the second is blue.

35) A magazine printer consecutively numbered the pages of a magazine, starting with 1 on the first page, 10 on the tenth page, etc. In numbering the gages, the printer printed a total of 195 digits.

Column A	Column B
The number of pages in the magazine	100

36)

Column A	Column B
The largest number that can be written by rearranging the digits in 381	The largest number that can be written by rearranging the digits in 279

37) A computer priced $147 includes 5% profit

Column A	Column B
$141	The original cost of the computer

IF YOU FINISH BEFORE TIME IS CALLED, YOU MAY CHECK YOUR WORK ON THIS SECTION ONLY. DO NOT TURN TO OTHER SECTION IN THE TEST. **STOP**

ISEE Upper Level Math Practice Test 1 Part 2 (Mathematics Achievement)

Total number of questions: 47

Total time for Part 1: 40 Minutes

You may NOT use a calculator on this part.

1) How is this number written in scientific notation?

$$0.00003379$$

A. 3.379×10^{-5}

B. 33.79×10^{6}

C. 0.3379×10^{-4}

D. 3379×10^{-8}

2) $|10 - (12 \div |1 - 5|)| = ?$

A. 7

B. -7

C. 5

D. -5

3) $(x + 4)(x + 5) =$

A. $x^2 + 9x + 20$

B. $2x + 12x + 12$

C. $x^2 + 25x + 10$

D. $x^2 + 12x + 35$

4) Find all values of x for which $6x^2 + 16x + 8 = 0$

A. $-\dfrac{3}{2}, -\dfrac{1}{2}$

B. $-\dfrac{2}{3}, -2$

C. $-2, -\dfrac{1}{4}$

D. $-\dfrac{2}{3}, \dfrac{1}{2}$

5) Which of the following graphs represents the compound inequality $-2 \le 2x - 4 < 8$?

A.

B.

C.

D.

6) $1 - 9 \div (4^2 \div 2) =$ ___

A. 6

B. $\frac{3}{4}$

C. $-\frac{1}{8}$

D. -2

7) A girl 200 *cm* tall, stands 460 *cm* from a lamp post at night. Her shadow from the light is 80 *cm* long. How high is the lamp post?

A. 440

B. 500

C. 900

D. 1350

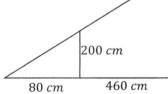

200 *cm*

80 *cm*   460 *cm*

8) Which graph corresponds to the following inequality?

$$-6y \leq 16x - 12$$

☐A.

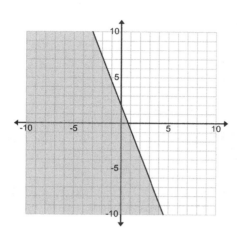

☐B.

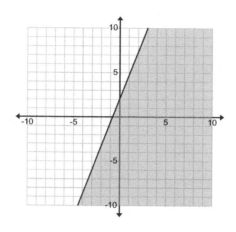

☐C.

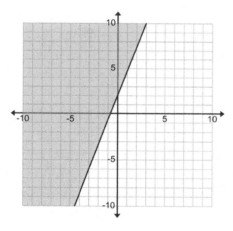

☐D.

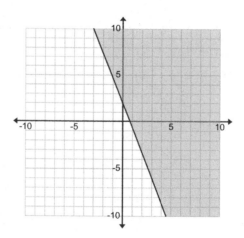

9) The ratio of boys to girls in a school is 3 : 5. If there are 600 students in a school, how many boys are in the school?

A. 200

B. 225

C. 300

D. 340

10) $90 \div \frac{1}{9} = ?$

A. 9.125

B. 10

C. 81

D. 810

11) The rectangle on the coordinate grid is translated 5 units down and 4 units to the left.

Which of the following describes this transformation?

A. $(x, y) \Rightarrow (x - 4, y + 5)$

B. $(x, y) \Rightarrow (x - 4, y - 5)$

C. $(x, y) \Rightarrow (x + 4, y + 5)$

D. $(x, y) \Rightarrow (x + 4, y - 5)$

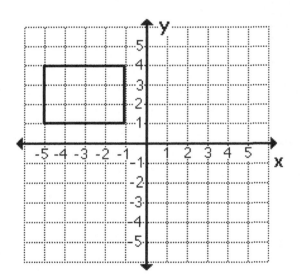

12) Find the area of a rectangle with a length of 148 feet and a width of 90 feet.

A. 13,320 $sq. ft$

B. 13,454 $sq. ft$

C. 13,404 $sq. ft$

D. 13,204 $sq. ft$

13) Which value of x makes the following inequality true?

$$\frac{4}{23} \le x < 25\%$$

A. 0.12

B. $\frac{5}{36}$

C. $\sqrt{0.044}$

D. 0.104

14) Which of the following could be the product of two consecutive prime numbers?

A. 2

B. 10

C. 14

D. 15

15) The perimeter of the trapezoid below is 36 cm. What is its area?

A. 576 cm^2

B. 70 cm^2

C. 48 cm^2

D. 24cm^2

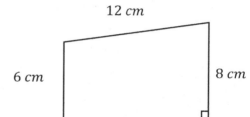

16) Emily lives $4\frac{1}{5}$ miles from where she works. When traveling to work, she walks to a bus stop $\frac{1}{2}$ of the way to catch a bus. How many miles away from her house is the bus stop?

A. $2\frac{1}{10}$ miles

B. $4\frac{3}{10}$ miles

C. $2\frac{3}{10}$ miles

D. $1\frac{3}{10}$ miles

17) If a vehicle is driven 40 miles on Monday, 45 miles on Tuesday, and 50 miles on Wednesday, what is the average number of miles driven each day?

A. 40 miles

B. 45 miles

C. 50 miles

D. 53 miles

18) Use the diagram below to answer the question.

Given the lengths of the base and diagonal of the rectangle below, what is the length of height h, in terms of s?

A. $s\sqrt{2}$

B. $2s\sqrt{2}$

C. $3s$

D. $3s^2$

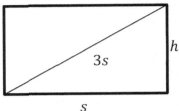

Use the chart below to answer the question.

Color	Number
White	40
Black	30
Beige	40

19) There are also purple marbles in the bag. Which of the following can NOT be the probability of randomly selecting a purple marble from the bag?

A. $\frac{1}{11}$

B. $\frac{1}{6}$

C. $\frac{2}{5}$

D. $\frac{1}{23}$

20) With an 23% discount, Ella was able to save $21.87 on a dress. What was the original price of the dress?

A. $89.92

B. $91.82

C. $95.09

D. $97.92

21) $\frac{8}{35}$ is equals to:

A. 0.28

B. 2.28

C. 0.028

D. 0.228

22) If 30% of A is 1,200, what is 12% of A?

A. 280

B. 480

C. 1,200

D. 1600

23) Simplify $\dfrac{\frac{1}{3} - \frac{x-5}{9}}{\frac{x^3}{3} - \frac{7}{3}}$

A. $\dfrac{7+x}{3x^3+21}$

B. $\dfrac{-7-x}{x^3-21}$

C. $\dfrac{7+x}{x^3-21}$

D. $\dfrac{7-x}{3x^3-21}$

24) If $(6.2 + 8.3 + 2.4) \times x = x$, then what is the value of x?

A. 0

B. $\dfrac{3}{10}$

C. -6

D. -12

25) Two dice are thrown simultaneously, what is the probability of getting a sum of 5 or 8?

A. $\dfrac{1}{3}$

B. $\dfrac{1}{4}$

C. $\dfrac{1}{16}$

D. $\dfrac{11}{36}$

26) If 7 garbage trucks can collect the trash of 38 homes in a day. How many trucks are needed to collect in 190 houses?

A. 20

B. 30

C. 35

D. 40

27) $78.56 \div 0.05 =$?

A. 15.712

B. 1,571.2

C. 157.12

D. 1.5712

28) In the following figure, AB is the diameter of the circle. What is the circumference of the circle?

A. 4π

B. 6π

C. 8π

D. 10π

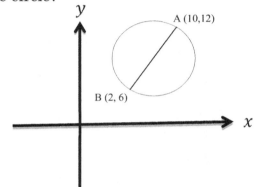

29) What is the value of x in the following equation?
$$2x^2 + 6 = 26$$

A. ± 4

B. $\pm \sqrt{9}$

C. $\pm \sqrt{10}$

D. ± 3

30) A circle has a diameter of 20 inches. What is its approximate area?

A. $314 \ inch^2$

B. $114 \ inch^2$

C. $74.00 \ inch^2$

D. $12.56 \ inch^2$

31) $6 \ days \ 20 \ hours \ 36 \ minutes - 4 \ days \ 12 \ hours \ 24 \ minutes =$?

A. $2 \ days \ 8 \ hours \ 12 \ minutes$

B. $1 \ days \ 8 \ hours \ 12 \ minutes$

C. $2 \ days \ 7 \ hours \ 14 \ minutes$

D. $1 \ days \ 7 \ hours \ 14 \ minutes$

32) The base of a right triangle is 4 feet, and the interior angles are 45 − 45 − 90. What is its area?

A. 2 *square feet*

B. 4 *square feet*

C. 8 *square feet*

D. 10 *square feet*

Use the following table to answer question below.

DANIEL'S BIRD-WATCHING PROJECT	
DAY	NUMBER OF RAPTORS SEEN
Monday	?
Tuesday	10
Wednesday	15
Thursday	13
Friday	6
MEAN	11

33) The above table shows the data Daniel collects while watching birds for one week. How many raptors did Daniel see on Monday?

A. 10

B. 11

C. 12

D. 13

34) A floppy disk shows 837,036 bytes free and 639,352 bytes used. If you delete a file of size 542,159 bytes and create a new file of size 499,986 bytes, how many free bytes will the floppy disk have?

A. 567,179 bytes

B. 671,525 bytes

C. 879,209 bytes

D. 899,209 bytes

35) Increased by 40%, the number 70 becomes:

A. 40

B. 98

C. 126

D. 130

36) If $12 + x^{\frac{1}{2}} = 24$, then what is the value of $5 \times x$?

A. 15

B. 60

C. 240

D. 720

37) Which equation represents the statement "twice the difference between 5 times H and 2 gives 35".

A. $\frac{5H + 2}{2} = 35$

B. $5(2H + 2) = 35$

C. $2(5H - 2) = 35$

D. $2\frac{5H}{2} = 35$

38) A circle is inscribed in a square, as shown below.
The area of the circle is $25\pi\ cm^2$ What is the area of the square?

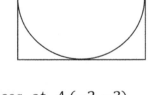

A. $10\ cm^2$

B. $26\ cm^2$

C. $48\ cm^2$

D. $100\ cm^2$

39) Triangle ABC is graphed on a coordinate grid with vertices at $A\ (-2, -3)$, $B\ (-4, 1)$ and $C\ (9, 7)$. Triangle ABC is reflected over x axes to create triangle $A'B'C'$.
Which order pair represents the coordinate of C'?

A. $(9, 7)$

B. $(-9, -7)$

C. $(9, -7)$

D. $(7, -9)$

40) Which set of ordered pairs represents y as a function of x?

A. $\{(5, -2), (5, 7), (9, -8), (4, -7)\}$

B. $\{(2, 2), (3, -9), (5, 8), (2, 7)\}$

C. $\{(9, 12), (8, 7), (6, 11), (8, 18)\}$

D. $\{(6, 1), (3, 1), (0, 5), (6, 1)\}$

41) The width of a box is one third of its length. The height of the box is one half of its width. If the length of the box is 24 cm, what is the volume of the box?

A. $80 cm^3$

B. $165\ cm^3$

C. $243\ cm^3$

D. $768 cm^3$

42) How many 4×4 squares can fit inside a rectangle with a height of 40 and width of 12?

A. 60

B. 50

C. 40

D. 30

43) David makes a weekly salary of \$230 plus 9% commission on his sales. What will his income be for a week in which he makes sales totaling \$1,200?

A. \$338

B. \$318

C. \$308

D. \$298

44) $5x^3y^2 + 4x^5y^3 - (6x^3y^2 - 4xy^5) =$ ___

A. $4x^5y^3$

B. $4x^5y^3 - x^3y^2 + 4xy^5$

C. $C.x^3y^2$

D. $4x^5y^3 - x^3y^2$

45) The radius of circle A is five times the radius of circle B. If the circumference of circle A is 20π, what is the area of circle B?

A. 3π

B. 4π

C. 6π

D. 12π

46) A square measures 8 inches on one side. By how much will the area be decreased if its length is increased by 5 inches and its width decreased by 4 inches.

A. 1 *sq decreased*

B. 3 *sq decreased*

C. 8 *sq decreased*

D. 12 *sq decreased*

47) If a box contains red and blue balls in ratio of $3:2$ red to blue, how many red balls are there if 80 blue balls are in the box?

A. 60

B. 80

C. 100

D. 120

IF YOU FINISH BEFORE TIME IS CALLED, YOU MAY CHECK YOUR WORK ON THIS SECTION.

STOP

ISEE UPPER LEVEL MATH PRACTICE TEST 2

2020 - 2021

Two Parts

- ▶ **Total number of questions:** 84
- ▶ **Part 1 (Calculator):** 37 questions
- ▶ **Part 2 (Calculator):** 47 questions
- ▶ **Total time for two parts:** 75 Minutes

ISEE Upper Level Practice Tests Answer Sheet

Remove (or photocopy) these answer sheets and use them to complete the practice tests.

ISEE Upper Level Practice Test			
Quantitative Reasoning		Mathematics Achievement	

1 (A)(B)(C)(D)	25 (A)(B)(C)(D)	1 (A)(B)(C)(D)	25 (A)(B)(C)(D)
2 (A)(B)(C)(D)	26 (A)(B)(C)(D)	2 (A)(B)(C)(D)	26 (A)(B)(C)(D)
3 (A)(B)(C)(D)	27 (A)(B)(C)(D)	3 (A)(B)(C)(D)	27 (A)(B)(C)(D)
4 (A)(B)(C)(D)	28 (A)(B)(C)(D)	4 (A)(B)(C)(D)	28 (A)(B)(C)(D)
5 (A)(B)(C)(D)	29 (A)(B)(C)(D)	5 (A)(B)(C)(D)	29 (A)(B)(C)(D)
6 (A)(B)(C)(D)	30 (A)(B)(C)(D)	6 (A)(B)(C)(D)	30 (A)(B)(C)(D)
7 (A)(B)(C)(D)	31 (A)(B)(C)(D)	7 (A)(B)(C)(D)	31 (A)(B)(C)(D)
8 (A)(B)(C)(D)	32 (A)(B)(C)(D)	8 (A)(B)(C)(D)	32 (A)(B)(C)(D)
9 (A)(B)(C)(D)	33 (A)(B)(C)(D)	9 (A)(B)(C)(D)	33 (A)(B)(C)(D)
10 (A)(B)(C)(D)	34 (A)(B)(C)(D)	10 (A)(B)(C)(D)	34 (A)(B)(C)(D)
11 (A)(B)(C)(D)	35 (A)(B)(C)(D)	11 (A)(B)(C)(D)	35 (A)(B)(C)(D)
12 (A)(B)(C)(D)	36 (A)(B)(C)(D)	12 (A)(B)(C)(D)	36 (A)(B)(C)(D)
13 (A)(B)(C)(D)	37 (A)(B)(C)(D)	13 (A)(B)(C)(D)	37 (A)(B)(C)(D)
14 (A)(B)(C)(D)		14 (A)(B)(C)(D)	38 (A)(B)(C)(D)
15 (A)(B)(C)(D)		15 (A)(B)(C)(D)	39 (A)(B)(C)(D)
16 (A)(B)(C)(D)		16 (A)(B)(C)(D)	40 (A)(B)(C)(D)
17 (A)(B)(C)(D)		17 (A)(B)(C)(D)	41 (A)(B)(C)(D)
18 (A)(B)(C)(D)		18 (A)(B)(C)(D)	42 (A)(B)(C)(D)
19 (A)(B)(C)(D)		19 (A)(B)(C)(D)	43 (A)(B)(C)(D)
20 (A)(B)(C)(D)		20 (A)(B)(C)(D)	44 (A)(B)(C)(D)
21 (A)(B)(C)(D)		21 (A)(B)(C)(D)	45 (A)(B)(C)(D)
22 (A)(B)(C)(D)		22 (A)(B)(C)(D)	46 (A)(B)(C)(D)
23 (A)(B)(C)(D)		23 (A)(B)(C)(D)	47 (A)(B)(C)(D)
24 (A)(B)(C)(D)		24 (A)(B)(C)(D)	

ISEE Upper Level Math Practice Test 2 Part 1 (Quantitative Reasoning)

Total number of questions: 37

Total time for Part 1: 35 Minutes

You may NOT use a calculator on this part.

1) How much greater is the value of $4x + 9$ than the value of $4x - 3$?

A. 8

B. 10

C. 12

D. 14

2) What is the prime factorization of 1,400?
A. $2 \times 2 \times 5 \times 5$

B. $2 \times 2 \times 2 \times 5 \times 5 \times 7$

C. 2×5

D. $2 \times 2 \times 2 \times 5 \times 7$

3) If 6 inches on a map represents an actual distance of 150 feet, then what actual distance does 20 inches on the map represent?
A. 180 feet

B. 200 feet

C. 250 feet

D. 500 feet

4) The circle graph below shows all Mr. Green's expenses for last month. If he spent $550 on his car, how much did he spend for his rent?
A. $675

B. $750

C. $780

D. $810

Mr. Green's monthly expenses

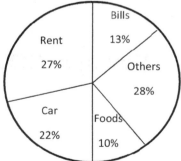

5) The area of a circle is less than 49π. Which of the following can be the circumference of the circle?
A. 10π

B. 14π

C. 24π

D. 32π

6) A basket contains 25 balls and the average weight of each of these balls is 35 g. The five heaviest balls have an average weight of 50 g each. If we remove the three heaviest balls from the basket, what is the average weight of the remaining balls?

A. 10 g

B. 20.25 g

C. 31.25 g

D. 35 g

7) If $f(x) = x^2 + 6$, what is the smallest possible value of $f(x)$?

A. 0

B. 5

C. 6

D. 7

8) Alice drives from her house to work at an average speed of 45 miles per hour and she drives at an average speed of 65 miles per hour when she was returning home. What was her minimum speed on the round trip in miles per hour?

A. 45

B. 58.5

C. 65

D. Cannot be determined

9) If the sum of the positive integers from 1 to n is 3,350, and the sum of the positive integers from $n + 1$ to $2n$ is 4,866, which of the following represents the sum of the positive integers from 1 to $2n$ inclusive?

A. 3,350

B. 4,866

C. 7,000

D. 8,216

10) Oscar purchased a new hat that was on sale for $8.34. The original price was $14.65. What percentage discount was the sale price?

A. 4.2%

B. 40.5%

C. 43%

D. 45%

11) Which of the following statements is correct, according to the graph below?

Number of Books Sold in a Bookstore

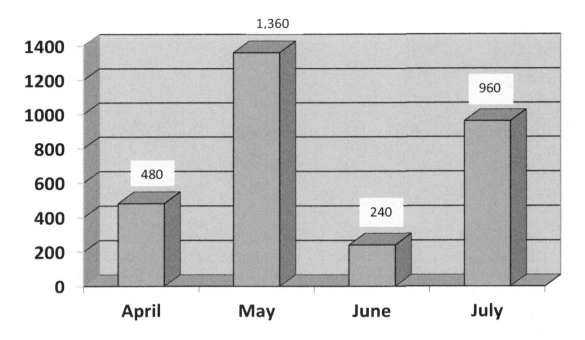

A. Number of books sold in April was twice the number of books sold in July.

B. Number of books sold in July was less than half the number of books sold in May.

C. Number of books sold in June was half the number of books sold in April.

D. Number of books sold in July was equal to the number of books sold in April plus the number of books sold in June.

12) List A consists of the numbers {2, 4, 9, 11, 16}, and list B consists of the numbers {5, 7, 13, 15, 18}.

2) If the two lists are combined, what is the median of the combined list?

A. 7

B. 8

C. 9

D. 10

13) A bag contains 19 balls: three green, five black, eight blue, a brown, a red and one white. If 17 balls are removed from the bag at random, what is the probability that a brown ball has been removed?

A. $\frac{1}{9}$

B. $\frac{1}{6}$

C. $\frac{16}{19}$

D. $\frac{17}{19}$

14) If Jim adds 150 stamps to his current stamp collection, the total number of stamps will be equal to $\frac{4}{3}$ the current number of stamps. If Jim adds 40% more stamps to the current collection, how many stamps will be in the collection?

A. 340

B. 453

C. 512

D. 630

15) If $x + y = 7$ and $x - y = 6$ then what is the value of $(x^2 - y^2)$?

A. 24

B. 42

C. 65

D. 90

16) The area of rectangle $ABCD$ is 108 square inches. If the length of the rectangle is three times the width, what is the perimeter of rectangle $ABCD$?

A. 48 inches

B. 67 inches

C. 76 inches

D. 86 inches

17) What's The ratio of boys and girls in a class is $7:4$. If there are 55 students in the class, how many more girls should be enrolled to make the ratio $1:1$?

A. 6

B. 10

C. 12

D. 15

18) The sum of 8 numbers is greater than 320 and less than 480. Which of the following could be the average (arithmetic mean) of the numbers?

A. 30

B. 35

C. 40

D. 45

19) A gas tank can hold 35 gallons when it is $\frac{5}{2}$ full. How many gallons does it contain when it is full?

A. 125

B. 62.5

C. 50

D. 14

20) Triangle ABC is similar to triangle ADE. What is the length of side EC?

A. 4

B. 10

C. 18

D. 45

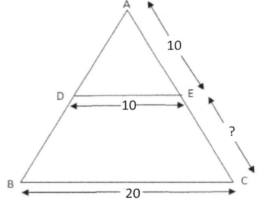

21) Which of the following expressions gives the value of b in terms of f, c, and z from the following equation?

$$f = [\frac{cz}{b}]^2$$

A. $b = fc^2z^2$

B. $b = \frac{cz}{\sqrt{f}}$

C. $b = \frac{\sqrt{f}}{cz}$

D. $b = [\frac{cz}{f}]^2$

Quantitative Comparisons

Direction: Questions 22 to 37 are Quantitative Comparisons Questions. Using the information provided in each question, compare the quantity in column A to the quantity in Column B. Choose on your answer sheet grid

A. if the quantity in Column A is greater

B. if the quantity in Column B is greater

C. if the two quantities are equal

D. if the relationship cannot be determined from the information given

22)

Column A	Column B
5^2	$\sqrt[3]{125}$

23)

Column A	Column B
7	$(52)^{\frac{1}{2}}$

24)

Column A	Column B
The average of $21, 29,$ and 37	28

25)

Column A	Column B
$16 \times 435 \times 25$	$19 \times 435 \times 22$

26) x is an integer

Column A	Column B
$-x$	$\dfrac{x}{3}$

27)

Column A	Column B
$(\frac{1}{4})^3$	4^{-3}

28) $3x + 7 > x - 1$

Column A	Column B
x	-7

29)

Column A	Column B
The greatest value of x in	The greatest value of x in
$8\,|3x - 2| = 16$	$8\,|3x - 2| = 16$

30) x is an integer

Column A	Column B
$(x)^5(x)^2$	$(x^5)^2$

31)

Column A	Column B
The probability that	The probability that
event x will occur.	event x will not occur.

32) The selling price of a sport jacket including 20% discount is $68.

Column A	Column B
	$80
Original price of the sport jacket	

33) $x^2 - 2x - 20 = 15$

Column A	Column B
x	5

34)

Column A	Column B
$(0.82)^{28}$	$(0.82)^{27}$

35)

Column A	Column B
The probability of rolling a 4 on a die and getting heads on a coin toss.	The probability of rolling an odd number on a die and picking a spade from a deck of 52 cards.

36)

Column A	Column B
$0.46	Sum of one quarter, three nickels, and three pennies

37) x is an odd integer, and y is an even integer. In a certain game an odd number is considered greater than an even number.

Column A	Column B
$x(x + y)$	$(x - y) - y^2$

IF YOU FINISH BEFORE TIME IS CALLED, YOU MAY CHECK YOUR WORK ON THIS SECTION.   **STOP**

ISEE Upper Level Math
Practice Test 2
Part 2
(Mathematics Achievement)

Total number of questions: 47

Total time for Part 1: 40 Minutes

You may NOT use a calculator on this part.

1) Which of the following points lies on the line $4x + 6y = 20$?

A. $(2, 1)$

B. $(-1, 3)$

C. $(-2, 2)$

D. $(2, 2)$

2) 5 less than twice a positive integer is 91. What is the integer?

A. 40

B. 41

C. 42

D. 48

3) If $\frac{|3+x|}{5} \le 8$, then which of the following is correct?

A. $-43 \le x \le 37$

B. $-43 \le x \le 32$

C. $-32 \le x \le 38$

D. $-32 \le x \le 32$

4) $\frac{1}{5b^2} + \frac{1}{5b} = \frac{1}{b^2}$, then $= $?

A. $-\frac{16}{5}$

B. 4

C. $-\frac{5}{16}$

D. 8

5) An angle is equal to one ninth of its supplement. What is the measure of that angle?

A. $18°$

B. $40°$

C. $60°$

D. $80°$

6) 1.3 is what percent of 26?

A. 1.3

B. 5

C. 18

D. 24

7) The cost, in thousands of dollars, of producing x thousands of textbooks is $C(x) = x^2 + 2x$. The revenue, also in thousands of dollars, is $R(x) = 40x$. find the profit or loss if 20 textbooks are produced. ($profit = revenue - cost$)

A. $2,160 profit

B. $360 profit

C. $2,160 loss

D. $360 loss

8) Simplify $7x^3y^3(2x^3y)^3 =$

A. $14x^4y^6$

B. $14x^8y^6$

C. $56x^{12}y^6$

D. $56x^8y^6$

9) Ella (E) is 5 years older than her friend Ava (A) who is 4 years younger than her sister Sofia (S). If E, A and S denote their ages, which one of the following represents the given information?

A. $\begin{cases} E = A + 5 \\ S = A - 4 \end{cases}$

B. $\begin{cases} E = A + 5 \\ A = S + 4 \end{cases}$

C. $\begin{cases} A = E + 5 \\ S = A - 4 \end{cases}$

D. $\begin{cases} E = A + 5 \\ A = S - 4 \end{cases}$

10) Right triangle ABC has two legs of lengths $4\ cm$ (AB) and $3\ cm$ (AC). What is the length of the third side (BC)?

A. $5\ cm$

B. $6\ cm$

C. $9\ cm$

D. $10\ cm$

11) Which is the longest time?

A. $24\ hours$

B. $1,520\ minutes$

C. $3\ days$

D. $4,200\ seconds$

12) A circle has a diameter of 10 inches. What is its approximate circumference?
A. 6.28 inches.

B. 25.12 inches.

C. 31.4 inches.

D. 35.12 inches.

13) Write 623 in expanded form, using exponents.
A. $(6 \times 10^3) + (2 \times 10^2) + (3 \times 10)$

B. $(6 \times 10^2) + (2 \times 10^1) - 5$

C. $(6 \times 10^2) + (2 \times 10^1) + 3$

D. $(6 \times 10^1) + (2 \times 10^2) + 3$

14) What is the area of an isosceles right triangle with hypotenuse that measures 8 cm?
A. 9 cm^2

B. 16 cm^2

C. $3\sqrt{2}$ cm^2

D. 64 cm^2

15) A company pays its writer \$5 for every 500 words written. How much will a writer earn for an article with 860 words?
A. \$12

B. \$5.6

C. \$8.6

D. \$10.7

16) A circular logo is enlarged to fit the lid of a jar. The new diameter is 20% larger than the original. By what percentage has the area of the logo increased?
A. 20%

B. 44%

C. 69%

D. 75%

17) $89.44 \div 0.05 = ?$

A. 17.888

B. 1,788.8

C. 178.88

D. 1.7888

18) What's the area of the non-shaded part of the following figure?

A. 225

B. 152

C. 40

D. 42

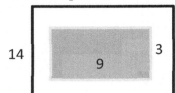

19) A bread recipe calls for $2\frac{1}{2}$ cups of flour. If you only have $1\frac{5}{4}$ cups, how much more flour is needed?

A. 1

B. $\frac{1}{2}$

C. 2

D. $\frac{1}{4}$

20) What is the maximum value for y if $y = -(x-2)^2 + 7$?

A. -7

B. -2

C. 2

D. 7

21) What is the solution of the following system of equations?
$$\begin{cases} -3x - y = -5 \\ 5x - 5y = 15 \end{cases}$$

A. $(-1, 2)$

B. $(2, -1)$

C. $(1, 4)$

D. $(4, -2)$

22) The equation of a line is given as: $y = 5x - 3$. Which of the following points does not lie on the line?

A. $(2,7)$

B. $(-2, -13)$

C. $(4, 21)$

D. $(-4, -23)$

23) The drivers at $G \& G$ trucking must report the mileage on their trucks each week. The mileage reading of Ed's vehicle was 52,806 at the beginning of one week, and 53,431 at the end of the same week. What was the total number of miles driven by Ed that week?

A. $515\ miles$

B. $525\ miles$

C. $625\ miles$

D. $658\ miles$

24) What is the area of an isosceles right triangle that has one leg that measures $4\ cm$?

A. $8\ cm^2$

B. $36\ cm^2$

C. $3\sqrt{2}\ cm^2$

D. $72\ cm^2$

25) Which of the following is a factor of both $x^2 - 5x + 6$ and $x^2 - 6x + 8$?
A. $(x - 2)$

B. $(x + 4)$

C. $(x + 2)$

D. $(x - 4)$

26)
$$
\begin{array}{r}
36\ \text{hr.}\ \ 38\ \text{min.}\\
-\ 23\ \text{hr.}\ \ 25\ \text{min.}\\
\hline
\end{array}
$$

A. $12\ hr.\,57\ min.$

B. $12\ hr.\,47\ min.$

C. $13\ hr.\,13\ min.$

D. $13\ hr.\,57\ min.$

27) $\frac{14}{26}$ is equal to:

A. 5.4

B. 0.54

C. 0.05

D. 0.5

28) If $x + y = 10$, what is the value of $9x + 9y$?

A. 192

B. 104

C. 90

D. 48

29) What is the number of cubic feet of soil needed for a flower box 2 feet long, 10 inches wide, and 2 feet deep?

A. 22 cubic feet

B. 12 cubic feet

C. $\frac{10}{3}$ cubic feet

D. 2 cubic feet

30) A car uses 20 gallons of gas to travel 460 miles. How many miles per gallon does the car use?

A. 23 miles per gallon

B. 32 miles per gallon

C. 30 miles per gallon

D. 34 miles per gallon

31) What is the reciprocal of $\frac{x^3}{15}$?

A. $\frac{15}{x^3} - 1$

B. $\frac{48}{x^3}$

C. $\frac{15}{x^3} + 1$

D. $\frac{15}{x^3}$

32) Karen is 9 years older than her sister Michelle, and Michelle is 4 years younger than her brother David. If the sum of their ages is 91, how old is Michelle?

A. 21

B. 26

C. 28

D. 29

33) Mario loaned Jett $1,400 at a yearly interest rate of 6%. After one year what is the interest owned on this loan?

A. $1,260

B. $140

C. $84

D. $30

34) Calculate the area of a parallelogram with a base of 3 feet and height of 3.2 feet.

A. 2.8 square feet

B. 4.2 square feet

C. 5.8 square feet

D. 9.6 square feet

35) Ellis just got hired for on-the-road sales and will travel about 2,500 miles a week during an 90-hour work week. If the time spent traveling is $\frac{5}{3}$ of his week, how many hours a week will he be on the road?

A. Ellis spends about 34 hours of his 90-hour work week on the road.

B. Ellis spends about 40 hours of his 90-hour work week on the road.

C. Ellis spends about 48 hours of his 90-hour work week on the road.

D. Ellis spends about 150 hours of his 90-hour work week on the road.

36) Given that $x = 0.5$ and $y = 5$, what is the value of $2x^2(y + 4)$?

A. 4.5

B. 8.2

C. 12.2

D. 14.2

37) What is the area of the shaded region if the diameter of the bigger circle is 14 inches and the diameter of the smaller circle is 10 inches.

A. $16\,\pi\ inch^2$

B. $24\,\pi\ inch^2$

C. $36\,\pi\ inch^2$

D. $80\,\pi\ inch^2$

38) A shirt costing $500 is discounted 25%. After a month, the shirt is discounted another 15%. Which of the following expressions can be used to find the selling price of the shirt?

A. $(500)\,(0.70)$

B. $(500) - 500\,(0.30)$

C. $(500)(0.15) - (500)\,(0.15)$

D. $(500)\,(0.75)\,(0.85)$

39) A tree 40 feet tall casts a shadow 18 feet long. Jack is 5 feet tall. How long is Jack's shadow?

A. $2.25\,ft$

B. $4\,ft$

C. $5.25\,ft$

D. $7\,ft$

40) In a school, the ratio of number of boys to girls is $7\!:\!3$. If the number of boys is 210, what is the total number of students in the school?

A. 300

B. 500

C. 540

D. 600

41) If x is 35% percent of 620, what is x?

A. 185

B. 217

C. 402

D. 720

42) How many square feet of tile is needed for a 19 $feet$ × 19 $feet$ room?

A. 72 square feet

B. 108 square feet

C. 361 square feet

D. 416 square feet

43) $(4x + 4)(x + 5) =$

A. $4x + 8$

B. $4x + 3x + 15$

C. $4x^2 + 24x + 20$

D. $4x^2 + 3$

44) If $x \blacksquare y = \sqrt{x^2 + y}$, what is the value of $4 \blacksquare 9$?

A. $\sqrt{126}$

B. 6

C. 5

D. 4

45) There are four equal tanks of water. If $\frac{2}{3}$ of a tank contains 200 liters of water, what is the capacity of the three tanks of water together?

A. 1,200 liters

B. 500 liters

C. 240 liters

D. 80 liters

46)

47) What is the result of the expression?

$$\begin{vmatrix} 3 & 6 \\ -1 & -3 \\ -5 & -1 \end{vmatrix} + \begin{vmatrix} 2 & -1 \\ 6 & 4 \\ 1 & 3 \end{vmatrix}$$

A. $\begin{vmatrix} 1 & -1 \\ 6 & 0 \\ 2 & 3 \end{vmatrix}$

B. $\begin{vmatrix} 3 & 7 \\ -1 & -3 \\ -5 & -1 \end{vmatrix}$

C. $\begin{vmatrix} 5 & 5 \\ 5 & 1 \\ -4 & 2 \end{vmatrix}$

D. $\begin{vmatrix} 5 & -3 \\ -6 & 1 \\ -10 & -3 \end{vmatrix}$

48) The average weight of 20 girls in a class is $55\,kg$ and the average weight of 35 boys in the same class is $70\,kg$. What is the average weight of all the 55 students in that class?

A. $60\,kg$

B. $61.28\,kg$

C. $64.54\,kg$

D. $65.9\,kg$

IF YOU FINISH BEFORE TIME IS CALLED, YOU MAY CHECK YOUR WORK ON THIS SECTION. **STOP**

ISEE UPPER LEVEL MATH PRACTICE TESTS ANSWER KEYS

Now, it's time to review your results to see where you went wrong and what areas you need to improve.

ISEE Upper Level Math Practice Test 1 Answer Key											
Quantitative Reasoning						Mathematics Achievement					
1	D	17	D	33	A	1	A	17	B	33	B
2	C	18	B	34	B	2	A	18	B	34	C
3	B	19	A	35	A	3	A	19	C	35	B
4	B	20	A	36	B	4	B	20	C	36	D
5	C	21	B	37	A	5	D	21	D	37	C
6	B	22	D			6	C	22	B	38	D
7	C	23	C			7	D	23	B	39	C
8	B	24	B			8	D	24	A	40	D
9	A	25	D			9	B	25	B	41	D
10	B	26	D			10	D	26	C	42	D
11	A	27	B			11	B	27	B	43	A
12	C	28	A			12	A	28	D	44	B
13	D	29	A			13	C	29	C	45	B
14	B	30	C			14	D	30	A	46	D
15	A	31	D			15	B	31	A	47	D
16	A	32	D			16	A	32	C		

ISEE Upper Level Math Practice Test 2 Answer Key

Quantitative Reasoning						Mathematics Achievement					
1	C	17	D	33	D	1	D	17	B	33	C
2	B	18	D	34	B	2	D	18	A	34	D
3	D	19	D	35	B	3	A	19	D	35	D
4	A	20	B	36	A	4	B	20	D	36	A
5	A	21	B	37	A	5	A	21	B	37	B
6	C	22	A			6	B	22	C	38	D
7	C	23	B			7	B	23	C	39	A
8	D	24	A			8	C	24	A	40	A
9	D	25	B			9	D	25	A	41	B
10	C	26	D			10	A	26	C	42	C
11	C	27	C			11	C	27	B	43	C
12	D	28	A			12	C	28	C	44	C
13	D	29	A			13	C	29	C	45	A
14	D	30	D			14	B	30	A	46	C
15	B	31	D			15	C	31	D	47	C
16	A	32	A			16	B	32	B		

ISEE Upper Level Math
Practice Tests
Answers and Explanations

ISEE UPPER LEVEL MATH PRACTICE TEST 3

QUANTITATIVE REASONING

1) Choice D is correct

$2y + 6 < 30 \rightarrow 2y < 30 - 6 \rightarrow 2y < 24 \rightarrow y < 12$, Only choice D (8) is less than 12.

2) Choice C is correct

A factor must divide evenly into its multiple. 16 cannot be a factor of 90 because 90 divided by 16 = 5.625

3) Choice B is correct

The diagonal of the square is 6 meters. Let x be the side.

Use Pythagorean Theorem: $a^2 + b^2 = c^2$

$x^2 + x^2 = 6^2 \Rightarrow 2x^2 = 6^2 \Rightarrow 2x^2 = 36 \Rightarrow x^2 = 18 \Rightarrow x = \sqrt{18}$

The area of the square is: $\sqrt{18} \times \sqrt{18} = 18 \, m^2$

4) Choice B is correct

To find what percent A is of B, divide A by B, then multiply that number by 100%:

$13.26 \div 58.56 = 0.2264 \times 100\% = 22.64\%$, This is approximately 22%.

5) Choice C is correct

$Probability = \dfrac{number\ of\ desired\ outcomes}{number\ of\ total\ outcomes}$

In this case, a desired outcome is selecting either a red or a yellow marble. Combine the number of red and yellow marbles: $9 + 6 = 15$, and divide this by the total number of marbles:

$7 + 9 + 6 = 22$. The probability is $\dfrac{15}{22}$.

6) Choice B is correct

Begin by calculating James's total earnings after 20 hours: $20 \, hours \times \$8.500 \, per \, hour = \$170$, Next, divide this total by Jacob's hourly rate to find the number of hours Jacob would need to work: $\$170 \div \$10.00 \, per \, hour = 17 \, hours$

7) Choice C is correct

The total cost of the phone call can be represented by the equation: $TC = \$5.00 + \$0.5x$, where x is the duration of the call after the first five minutes. In this case, $x = 30$. Substitute the known values into the equation and solve:

$TC = \$5.00 + \$0.5 \times 30 \rightarrow TC = \$5.00 + \$15.00$, $TC = \$20.00$

8) Choice B is correct

Let b be the amount of time Alec can do the job, then,

$$\frac{1}{a} + \frac{1}{b} = \frac{1}{50} \rightarrow \frac{1}{150} + \frac{1}{b} = \frac{1}{50} \rightarrow \frac{1}{b} = \frac{1}{50} - \frac{1}{150} = \frac{2}{150} = \frac{1}{75}$$

Then: $b = 75$ minutes

9) Choice A is correct

First, find the number. Let x be the number. Write the equation and solve for x. 150% of a number is 75, then: $1.5 \times x = 75 \rightarrow x = 75 \div 1.5 = 50$, 90% of 50 is: $0.9 \times 50 = 45$

10) Choice B is correct

The length of MN is equal to: $4x + 6x = 10x$, Then: $10x = 50 \rightarrow x = \frac{50}{10} = 5$

The length of ON is equal to: $6x = 6 \times 5 = 30\ cm$

11) Choice A is correct

The distance on the map is proportional to the actual distance between the two cities. Use the information to set up a proportion and then solve for the unknown number of actual miles: $\frac{14\ miles}{\frac{1}{2}\ inches} = \frac{x\ miles}{18\ inc}$, Cross multiply and simplify to solve for the x:

$$\frac{14 \times 18}{\frac{1}{2}} = x\ miles \rightarrow \frac{252}{\frac{1}{2}} = 252 \times 2 = 504\ miles$$

12) Choice C is correct

Number of Mathematics books: $0.35 \times 700 = 245$

Number of English books: $0.13 \times 700 = 91$

Product of number of Mathematics and number of English books: $245 \times 91 = 22,295$

13) Choice D is correct

If each book weighs $\frac{1}{6}$ pound, then $1\ pound = 6\ books$. To find the number of books in 60 pounds, simply multiply this 6 by 60: $60 \times 6 = 360$

14) Choice B is correct

Write equations based on the information provided in the question:

$Emily = Daniel, Emily = 6\ Claire, Daniel = 15 + Claire$

$Emily = Daniel \rightarrow Emily = 15 + Claire$

$Emily = 6\ Claire \rightarrow 6\ Claire = 15 + Claire \rightarrow 6\ Claire - Claire = 15$

$5\ Claire = 15, Claire = 3$

15) Choice A is correct

Begin by examining the sequence to find the pattern. The difference between 3 and 6 is 3; moving from 6 to 11 requires 5 to be added; moving from 11 to 18 requires 7 to be added. The pattern emerges here — adding by consecutive odd integers. The 5th term is equal to $18 + 9 = 27$, and the 6th term is equal to $27 + 11 = 38$.

16) Choice A is correct

The general slope-intercept form of the equation of a line is $y = mx + b$, where m is the slope and b is the y-intercept. By substitution of the given point and given slope: $-3 = (3)(2) + b$, So, $b = -3 - 6 = -9$, and the required equation is $y = 3x - 9$.

17) Choice D is correct

Multiplying each side of $-4x - y = 8$ by 2 gives $-8x - 2y = 16$. Adding each side of $-8x - 2y = 16$ to the corresponding side of $8x + 6y = 20$ gives $4y = 36$ or $y = 9$. Finally, substituting 9 for y in $8x + 6y = 20$ gives $8x + 6(9) = 20$ or $x = -\frac{17}{4}$.

18) Choice B is correct

The question is this: 530.20 is what percent of 631? Use percent formula:

$Part = \frac{percent}{100} \times whole,$

$530.20 = \frac{percent}{100} \times 631 \rightarrow 530.20 = \frac{percent \times 631}{100} \rightarrow 53020 = percent \times 631$

Then, Percent $= \frac{53020}{631} = 84.02$

530.20 is 84% of 631. Therefore, the discount is: $100\% - 84\% = 16\%$

19) Choice A is correct

Let x be the number. Write the equation and solve for x. $\frac{3}{2} \times 20 = \frac{5}{2}x \rightarrow \frac{3 \times 20}{2} = \frac{5x}{2}$, use cross multiplication to solve for x. $2 \times 60 = 5x \times 2 \Rightarrow 120 = 10x \Rightarrow x = 12$

20) Choice A is correct

Let's choose $100 for the sales of the supermarket. If the sales increases by 11 percent in April, the final amount of sales at the end of April will be $100 + (11\%) \times (\$100) = \$111$.

If sales then decreased by 11 percent in May, the final amount of sales at the end of August will be $\$111 - (\$111) \times (11\%) = \$98.79$

The final sales of $98 is 98% of the original price of $100. Therefore, the sales decreased by 2% overall.

21) Choice B is correct

The time it takes to drive from city A to city B is: $\frac{2700}{74} = 36.48$

22) Choice D is correct

First find the value of x in the equation: $x^2 = 8 \rightarrow x = \sqrt{8}$, or $x = -\sqrt{8}$. Since $\sqrt{8}$ is bigger than 2 and $-\sqrt{8}$ is smaller than 2. Then: the relationship cannot be determined from the information given.

23) Choice C is correct

Find the value of the functions in each column.

Column A: $4 \diamondsuit 5 = 4^2 - 2(4)(5) + 5^2 = 16 - 40 + 25 = 1$,

Column B: $5 \diamondsuit 4 = 5^2 - 2(5)(4) + 4^2 = 25 - 40 + 16 = 1$

The two quantities are equal

24) Choice B is correct

Column A: $\frac{16+22+24+36+4}{5} = \frac{138}{5} = 27.6$, Column B: $\frac{28+33+38+42}{4} = \frac{141}{4} = 35.25$

25) Choice D is correct

Let's plug in a value for x and compare the values in both columns.

$x = -1$

Column A: $\frac{1}{x+1} = \frac{1}{-1+1} = \frac{1}{0}$, this is undefined.

Column B: $\frac{2}{x+2} = \frac{2}{-1+2} = \frac{2}{1} = 2$

The relationship cannot be determined from the information given

26) Choice D is correct

First, solve the expression for x. $\frac{x}{3} = y^2 \rightarrow x = 3y^2$

Plug in different values for y and find the values of x.

Let's choose $y = 0 \rightarrow x = 3y^2 \rightarrow x = 3(0)^2 = 0$

The values in Column A and B are equal.

Now, let's choose $y = 1 \rightarrow x = 3y^2 \rightarrow x = 3(1)^2 = 3$

Column A is greater. So, the relationship cannot be determined from the information given.

27) Choice B is correct

Column A: $2x^3 - 3x - 2 = 2(1)^3 - 3(1) - 2 = 2 - 3 - 2 = -3$

Column B: $3x^2 - 2x - 3 = 3(1)^2 - 2(1) - 3 = 3 - 2 - 3 = -2$

28) Choice A is correct

Column A: $8 + 14(7 - 3) = 64$, Column B: $14 + 8(7 - 3) = 46$

29) Choice A is correct

First find the value of x. $\frac{x}{36} = \frac{3}{4} \rightarrow 4x = 3 \times 36 = 108 \rightarrow x = \frac{10}{4} = 27$

Column A: $\frac{9}{x} = \frac{9}{27} = \frac{1}{3}$

30) Choice C is correct

First convert hours to minutes. 3 hours 10 minutes $= 3 \times 60 + 10 = 190$ minutes.

Machine D makes b rolls of steel in 38 minutes. So, it makes 5 sets of b in 190 minutes.

$190 \div 38 = 5$ sets of b.

Machine E operates for 5 hours, making b rolls per hour. So, it makes a total of $5b$ rolls.

The two quantities are equal

31) Choice D is correct

$-1 < y < 4$,   Let's choose some values for y. $y = 1$

Column A: $\frac{y}{2} = \frac{1}{2}$, Column B: $\frac{2}{y} = \frac{2}{1} = 2$, In this case, column B is bigger.

$y = 3$, Column A: $\frac{y}{2} = \frac{3}{2}$, Column B: $\frac{2}{y} = \frac{2}{3}$

In this case, Column A is bigger. So, the relationship cannot be determined from the information given.

32) Choice D is correct

$\frac{a}{b} = \frac{c}{d}$, here there are two equal fractions. Let's choose some values for these variables. $\frac{1}{2} = \frac{2}{4}$

In this case, Column A is 3 $(1 + 2)$ and Column B is 6 $(2 + 4)$. Since, we can change the positions of these variables (for example put 2 for a and 4 for b), the relationship cannot be determined from the information given.

33) Choice A is correct

The ratio of boys to girls in a class is 5 to 7. Therefore, ratio of boys to the entire class is 5 out of 12. $\frac{5}{12} > \frac{1}{3}$

34) Choice B is correct

There are 9 marbles in the jar. Let's calculate each probability individually:

The probability that the first marble is blue $= \frac{4}{9}$

The probability that the second marble is blue $= \frac{3}{8}$

Column A: The probability that both marbles are blue $= \frac{4}{9} \times \frac{3}{8} = \frac{12}{72} = \frac{1}{6}$

The probability that the first marble is green $= \frac{5}{9}$

The probability that the second marble is blue $= \frac{4}{8} = \frac{1}{2}$

Column B: The probability that the first marbles is green, but the second is blue $= \frac{5}{9} \times \frac{1}{2} = \frac{5}{18}$

Column B is greater. $\frac{5}{18} > \frac{1}{6}$.

35) Choice A is correct

First, let's find the number of digits when the printer prints 100 pages.

If there are 2 digits in each page and the printer prints 100 pages, then, there will be 200 digits. $100 \times 2 = 200$

However, we know that pages $1 - 9$ have only one digit each, so we must subtract 9 from this total: $200 - 9 = 191$. We also know that the number 100th has three digits not two. So, we must add 1 digit to this total: $191 + 1 = 192$.

It is given that 195 digits were printed, and we know that 100 pages results in 192 digits total, so there must be 101 total pages in the magazine. Column A is greater.

36) Choice B is correct

Column A: The largest number that can be written by rearranging the digits in $381 = 831$

Column B: The largest number that can be written by rearranging the digits in $279 = 972$

37) Choice A is correct

The computer priced \$147 includes 5% profit. Let x be the original cost of the computer.   Then: $x + 5\%$ of $x = 147 \rightarrow x + 0.05x = 147 \rightarrow 1.05x = 147 \rightarrow x = \frac{147}{1.05} = \$140$

Column A is bigger.

ISEE UPPER LEVEL MATH PRACTICE TEST 1

MATHEMATICS ACHIEVEMENT

1) Choice A is correct.

$0.00003379 = \frac{3.379}{100,000} \Rightarrow 3.379 \times 10^{-5}$

2) Choice A is correct

$|10 - (12 \div |1 - 5|)| = |10 - (12 \div |-4|)| = |10 - (12 \div 4)| = |10 - 3| = |7| = 7$

3) Choice A is correct

Use FOIL (First, Out, In, Last) method.

$(x + 4)(x + 5) = x^2 + 5x + 4x + 20 = x^2 + 9x + 20$

4) Choice B is correct

Use quadratic formula: $ax^2 + bx + c = 0$

$x_{1,2} = \frac{-b \pm \sqrt{b^2 - 4ac}}{2a}$

$6x^2 + 16x + 8 \quad \Rightarrow \quad$ then: $a = 6, b = 16$ and $c = 8$

$x = \frac{-16 + \sqrt{16^2 - 4 \times 6 \times 8}}{2 \times 6} = -\frac{2}{3}$

$x = \frac{-16 - \sqrt{16^2 - 4 \times 6 \times 8}}{2 \times 6} = -2$

5) Choice D is correct

Solve for x. $-2 \leq 2x - 4 < 8 \Rightarrow$ (add 4 all sides) $-2 + 4 \leq 2x - 4 + 4 < 8 + 4 \Rightarrow 2 \leq 2x < 12 \Rightarrow$ (divide all sides by 2) $1 \leq x < 6$. x is between 1 and 6. Choice D represent this inequality.

6) Choice C is correct

Simplify: $1 - 9 \div (4^2 \div 2) = -\frac{1}{8}$

7) Choice D is correct.

Write the proportion and solve for missing side.

$$\frac{\text{Smaller triangle height}}{\text{Smaller triangle base}} = \frac{\text{Bigger triangle height}}{\text{Bigger triangle base}} \Rightarrow \frac{80cm}{200cm} = \frac{80+460cm}{x} \Rightarrow x = 1,350 \ cm$$

8) Choice D is correct.

First, notice that the line $-6y = 16x - 12$ has a negative slope. (graph the line and see that the line has a negative slope) Then, Choices B and C that show the line with positive slope are incorrect.

Now, choose the testing point $(0,0)$ and plug in the values of x and y in the inequality. $(0,0) \rightarrow -6y \le 16x - 12 \rightarrow -6(0) \le 16(0) - 12 \rightarrow 0 \le -12$

Number 0 is not less than -12. Then, the testing point $(0,0)$ is not in the solution section. Therefore, choices A and B are incorrect. (notice that both graphs of A and B show the testing point $(0,0)$ in the solution section.)

Only choice D represents the inequality $-6y \le 16x - 12$

9) Choice B is correct

The ratio of boy to girls is $3:5$. Therefore, there are 3 boys out of 8 students. To find the answer, first divide the total number of students by 8, then multiply the result by 3.

$600 \div 8 = 75 \Rightarrow 75 \times 3 = 225$

10) Choice D is correct

$90 \div \frac{1}{9} = 90 \times 9 = 810$

11) Choice B is correct.

Translated 5 units down and 4 units to the left means: $(x.y) \Rightarrow (x - 4, y - 5)$

12) Choice A is correct

Area of a rectangle $= width \times height$, $Area = 148 \times 90 = 1,3320 \ sq. ft$

13) Choice C is correct.

$\frac{4}{23} = 0.173$ and $25\% = 0.25$ therefore x should be between 0.173 and 0.25. Only choice B $(\sqrt{0.044}) = 0.20$ is between 0.173 and 0.25.

14) Choice D is correct

Some of prime numbers are: $2, 3, 5, 7, 11, 13$. Find the product of two consecutive prime numbers: $2 \times 3 = 6$ (not in the options), $3 \times 5 = 15$ (bingo!), $5 \times 7 = 35$ (not in the options)

15) Choice B is correct

The perimeter of the trapezoid is $36 \ cm$.

Therefore, the missing side (height) is $= 36 - 8 - 12 - 6 = 10 \ cm$

Area of a trapezoid: $A = \frac{1}{2} h (b_1 + b_2) = \frac{1}{2} (10)(6 + 8) = 70 \ cm^2$

16) Choice A is correct

$\frac{1}{2}$ of the distance $4\frac{1}{5}$ miles is: $\frac{1}{2} \times 4\frac{1}{5} = \frac{1}{2} \times \frac{21}{5} = \frac{21}{10}$, Converting $\frac{21}{10}$ to a mixed number gives: $\frac{21}{10} = 2\frac{1}{10}$

17) Choice B is correct

$average = \frac{sum}{total} = \frac{40 + 45 + 50}{3} = \frac{135}{3} = 45 \ miles$

18) Choice B is correct

Use Pythagorean theorem: $a^2 + b^2 = c^2 \rightarrow s^2 + h^2 = (3s)^2 \rightarrow s^2 + h^2 = 9s^2$

Subtracting s^2 from both sides gives: $h^2 = 8s^2$

Square roots of both sides: $h = \sqrt{8s^2} = \sqrt{4 \times 2 \times s^2} = \sqrt{4} \times \sqrt{2} \times \sqrt{s^2} = 2 \times s \times \sqrt{2} = 2s\sqrt{2}$

19) Choice C is correct

Let x be the number of purple marbles. Let's review the choices provided:

A. $\frac{1}{11}$, if the probability of choosing a purple marble is one out of ten, then:

$Probability = \frac{number\ of\ desired\ outcomes}{number\ of\ total\ outcomes} = \frac{x}{40+40+30+} = \frac{1}{11}$

Use cross multiplication and solve for x. $11x = 110 + x \rightarrow 10x = 110 \rightarrow x = 11$

Since, number of purple marbles can be 9, then, choice be the probability of randomly selecting a purple marble from the bag.

Use same method for other choices.

B. $\frac{1}{6}$

$\frac{x}{40+40+30+x} = \frac{1}{6} \rightarrow 6x = 110 + x \rightarrow 5x = 110 \rightarrow x = 22$

C. $\frac{2}{5}$

$\frac{x}{40+40+30+} = \frac{2}{5} \rightarrow 5x = 220 + 2x \rightarrow 3x = 220 \rightarrow x = 73.3$

D. $\frac{1}{23}$

$\frac{x}{40+40+30+x} = \frac{1}{23} \rightarrow 23x = 110 + x \rightarrow 22x = 110 \rightarrow x = 5$

Number of purple marbles cannot be a decimal.

20) Choice C is correct

Let x be the original price of the dress. Then: 23% of $x = 21.87$

$x = \frac{23}{100}x = 21.87$, $x = \frac{100 \times 21.87}{23} \cong 95.09$

21) Choice D is correct

$\frac{8}{35} = 0.228$

22) Choice B is correct

30% of A is $1,200$ Then: $0.3A = 1,200 \rightarrow A = \frac{1,200}{0.3} = 4,000$

12% of $4,000$ is: $0.12 \times 4,000 = 480$

23) Choice D is correct

Simplify:

$$\frac{\frac{1}{3} - \frac{x-4}{9}}{\frac{x^3}{3} - \frac{7}{3}} = \frac{\frac{1}{3} - \frac{x-4}{9}}{\frac{x^3 - 7}{3}} = \frac{3(\frac{1}{3} - \frac{x-4}{9})}{x^3 - 7}$$

$\Rightarrow$ Simplify: $\dfrac{1}{3} - \dfrac{x-4}{9} = \dfrac{7-x}{9}$

Then: $\dfrac{3(\frac{7-x}{9})}{x^3 - 7} = \dfrac{\frac{7-x}{3}}{x^3 - 7} = \dfrac{7-x}{3(x^3 - 7)} = \dfrac{7-x}{3x^3 - 21}$

24) Choice A is correct

$(6.2 + 8.3 + 2.4) \times x = x$, $16.9x = x$, Then: $x = 0$

25) Choice B is correct

For sum of 5: $(1 \ \& \ 4)$ *and* $(4 \ \& \ 1), (2 \ \& \ 3)$ and $(3 \ \& \ 2)$, therefore we have 4 options.

For sum of 8: $(5 \ \& \ 3)$ *and* $(3 \ \& \ 5), (4 \ \& \ 4)$ and $(2 \ \& \ 6), (6 \ \& \ 2)$, we have 5 options. To get a sum of 5 or 8 for two dice: $4 + 5 = 9$

Since, we have $6 \times 6 = 36$ total number of options, the probability of getting a sum of 5 and 8 is 9 out of 36 or $\dfrac{9}{36} = \dfrac{1}{4}$

26) Choice C is correct

Write a proportion and solve. $\dfrac{7}{38} = \dfrac{x}{190} \rightarrow x = \dfrac{7 \times 190}{38} = 35$

27) Choice B is correct

$78.56 \div 0.05 = 1,571.2$

28) Choice D is correct

The distance of A to B on the coordinate plane is: $\sqrt{(x_1 - x_2)^2 + (y_1 - y_2)^2} = \sqrt{(2-10)^2 + (6-12)^2} = \sqrt{8^2 + 6^2}, = \sqrt{64 + 36} = \sqrt{100} = 10$

The diameter of the circle is 10 and the radius of the circle is 5. Then: the circumference of the circle is: $2\pi r = 2\pi(5) = 10\pi$

29) Choice C is correct

$2x^2 + 6 = 26 \to 2x^2 = 20 \to x^2 = 10 \to x = \pm\sqrt{10}$

30) Choice A is correct

Diameter = 20, then: Radius = 10, Area of a circle = $\pi r^2 \Rightarrow A = 3.14(10)^2 = 314$

31) Choice A is correct

$6\ days\ 20\ hours\ 36\ minutes - 4\ days\ 12\ hours\ 24\ minutes = 2\ days\ 8\ hours\ 12\ minutes$

32) Choice C is correct

Formula of triangle area $= \frac{1}{2}(base \times height)$. Since the angles are $45 - 45 - 90$, then this is an isosceles triangle, meaning that the base and height of the triangle are equal.

$Triangle\ area = \frac{1}{2}(base \times height) = \frac{1}{2}(4 \times 4) = 8\ square\ feet$

33) Choice B is correct

The mean of the data is 11. Then: $\frac{x+10+15+13}{5} = 11 \to x + 44 = 55 \to x = 55 - 44 = 11$

34) Choice C is correct

The difference of the file added, and the file deleted is:

$542,159 - 499,986 = 42,173$ bytes

$837,036 + 42,173 = 879,209$ bytes

35) Choice B is correct

$40\%\ of\ 70 = 28 \to 70 + 28 = 98$

36) Choice D is correct

$x^{\frac{1}{2}}$ equals to the root of x. Then: $12 + x^{\frac{1}{2}} = 24 \to 12 + \sqrt{x} = 24 \to \sqrt{x} = 12 \to x = 144$

$x = 144$ and $5 \times x$ equals: $5 \times 144 = 720$

37) Choice C is correct

Only choice C represents the statement "twice the difference between 5 times H and 2 gives 35". $2(5H - 2) = 35$

38) Choice D is correct

The area of the circle is $25\pi\ cm^2$, then, its diameter is $10cm$. $area\ of\ a\ circle = \pi r^2 = 25\pi \rightarrow r^2 = 25 \rightarrow r = 5\ cm$ .Radius of the circle is 5 and diameter is twice of it, 10.One side of the square equals to the diameter of the circle. Then:$Area\ of\ square = side \times side = 10 \times 10 = 100\ cm^2$

39) Choice C is correct.

When a point is reflected over x axes, the (y) coordinate of that point changes to $(-y)$ while its x coordinate remains the same. C $(9,7) \rightarrow$ C' $(9,-7)$

40) Choice D is correct.

A set of ordered pairs represents y as a function of x if: $x_1 = x_2 \rightarrow y_1 = y_2$

In choice A: $(5,-2)$ and $(5,7)$ are ordered pairs with same x and different y, therefore y isn't a function of x.

In choice B: $(2,2)$ and $(2,7)$ are ordered pairs with same x and different y, therefore y isn't a function of x.

In choice C: $(8,7)$ and $(8,18)$ are ordered pairs with same x and different y, therefore y isn't a function of x.

41) Choice D is correct

If the length of the box is $24\ cm$, then the width of the box is one third of it, $8\ cm$, and the height of the box is $4\ cm$ (half of the width). The volume of the box is:

$V = length \times width \times height = (24)(8)(4) = 768\ cm^3$

42) Choice D is correct

Number of squares equal to: $\frac{40 \times 12}{4 \times 4} = 10 \times 3 = 30$

43) Choice A is correct

David's weekly salary is \$230 plus 9% of \$1,200. Then: $9\%\ of\ 1,200 = 0.09 \times 1,200 = 108$

$230 + 108 = 338$

44) Choice B is correct

$5x^3y^2 + 4x^5y^3 - (6x^3y^2 - 4xy^5) = 5x^3y^2 + 4x^5y^3 - 6x^3y^2 + 4xy^5) = 4x^5y^3 - x^3y^2 + 4xy^5$

45) Choice B is correct

Let P be circumference of circle A, then; $2\pi r_A = 20\pi \rightarrow r_A = 10$

$r_A = 5r_B \rightarrow r_B = \frac{10}{5} = 2 \rightarrow$ Area of circle B is; $\pi r_B^2 = 4\pi$

46) Choice D is correct

The area of the square is 64 square inches. $Area\ of\ square = side \times side = 8 \times 8 = 64\ inch^2$

The length of the square is increased by 5 inches and its width decreased by 4 inches. Then, its area equals: *Area of rectangle* = *width* × *length* = $13 \times 4 = 52$

The area of the square will be decreased by 12 square inches. $64 - 52 = 12$

47) Choice D is correct

Write a proportion and solve. $\frac{3}{2} = \frac{x}{80}$, Use cross multiplication: $2x = 240 \rightarrow x = 120$

ISEE UPPER LEVEL MATH PRACTICE TEST 2

QUANTITATIVE REASONING

1) Choice C is correct

$(4x + 9) - (4x - 3) = 4x - 4x + 9 + 3 = 12$

2) Choice B is correct

Find the value of each choice:

$2 \times 2 \times 5 \times 5 = 100$

$2 \times 2 \times 2 \times 5 \times 5 \times 7 = 1,400$

$2 \times 7 = 14$

$2 \times 2 \times 2 \times 5 \times 7 = 280$

3) Choice D is correct

Write a proportion and solve. $\frac{6in}{150feet} = \frac{20in}{x} \rightarrow x = \frac{150 \times 20}{6} = 500 \ feet$

4) Choice A is correct

Let x be all expenses, then $\frac{22}{100}x = \$550 \rightarrow x = \frac{100 \times \$550}{22} = \$2,500$

He spent for his rent: $\frac{27}{100} \times \$2,500 = \$675$

5) Choice A is correct

Area of the circle is less than 14π. Use the formula of areas of circles. *Area* $= \pi r^2 \Rightarrow 49\pi > \pi r^2 \Rightarrow 49 > r^2 \Rightarrow r < 7$. Radius of the circle is less than 7. Let's put 7 for the radius. Now, use the circumference formula: *Circumference* $= 2\pi r = 2\pi(7) = 14\pi$. Since the radius of the circle is less than 7. Then, the circumference of the circle must be less than 14π. Only choice A is less than 14π

6) Choice C is correct

Recall that the formula for the average is: $Average = \frac{sum\ of\ data}{number\ of\ data}$

First, compute the total weight of all balls in the basket: $35g = \frac{total\ weigh}{25\ balls}$

$35g \times 25 = total\ weight = 875g$

Next, find the total weight of the 5 largest marbles: $50g = \frac{total\ weigh}{5\ marbles}$

$50\ g \times 5 = total\ weight = 250\ g$

The total weight of the heaviest balls is $250\ g$. Then, the total weight of the remaining 20 balls is $625g$. $875\ g - 250\ g = 625\ g$.

The average weight of the remaining balls: $Average = \frac{625\ g}{20\ marbles} = 31.25g$ per ball

7) Choice C is correct

The smallest possible value of $f(x)$ will occur when $x = 0$. Since x^2 is always positive, any positive or negative value of x will make the value of $f(x)$ greater than 6. Substitute 0 for x and evaluate the expression: $f(0) = (0)^2 + 6 = 6$

8) Choice D is correct

There is not enough information to determine the answer of the question. An average speed represents a distance divided by time and it does not provide information about the speed at specific time. Alice could drove exactly 45 miles per hour from start to finish, or she could drive 65 miles per hour for half of distance and 45 miles per hour for the other half.

9) Choice D is correct

There are 2 sets of values, one set from 1 to n, and the other set from $n + 1$ to $2n$. Since the second set begins immediately after the first set, the two sets can be combined. The sum of the positive integers from 1 to $2n$ inclusive is equal to the sum of the positive integers from 1 to n plus the sum of the positive integers from $n + 1$ to $2n$: $3,350 + 4,866 = 8,216$

10) Choice C is correct

The percentage discount is the reduction in price divided by the original price. The difference between original price and sale price is: $\$14.65 - \$8.34 = \$6.31$

The percentage discount is this difference divided by the original price: $\$6.31 \div \$14.65 \cong 0.43$

Convert the decimal to a percentage by multiplying by 100%: $0.43 \times 100\% = 43\%$

11) Choice C is correct

Let's review the choices provided:

A. Number of books sold in April is: 480

Number of books sold in July is: $960 \rightarrow \frac{480}{960} = \frac{48}{96} = \frac{1}{2}$

B. number of books sold in July is: 960

Half the number of books sold in May is: $\frac{1,360}{2} = 680 \rightarrow 960 > 680$

C. number of books sold in June is: 240

Half the number of books sold in April is: $\frac{480}{2} = 240 \rightarrow 240 = 240$

D. $480 + 240 = 720 < 960$

Only choice C is correct.

12) Choice D is correct

The median of a set of data is the value located in the middle of the data set. Combine the 2 sets provided, and organize them in ascending order: $\{2, 4, 5, 7, 9, 11, 13, 15, 16, 18\}$

Since there are an even number of items in the resulting list, the median is the average of the two middle numbers. $Median = (9 + 11) \div 2 = 10$

13) Choice D is correct

If 17 balls are removed from the bag at random, there will be one ball in the bag. The probability of choosing a brown ball is 1 out of 19. Therefore, the probability of not choosing a brown ball is 17 out of 19 and the probability of having not a brown ball after removing 17 balls is the same.

14) Choice D is correct

Let x be the number of current stamps in the collection. Then: $\frac{4}{3}x - x = 150 \rightarrow$ $\frac{1}{3}x = 150 \rightarrow x = 450$, 40% more of 450 is: $450 + 0.40 \times 450 = 450 + 180 = 630$

15) Choice B is correct

$(x^2 - y^2) = (x - y)(x + y)$, Then: $x^2 - y^2 = 7 \times 6 = 42$

16) Choice A is correct

The formula for the area of a rectangle is: $Area = Width \times Length$

It is given that $L = 3W$ and that $A = 108$. Substitute the given values into our equation and solve for W: $108 = w \times 3w \rightarrow 108 = 3w^2 \rightarrow w^2 = 36 \rightarrow w = 6$

It is given that $L = 3W$, therefore, $L = 3 \times 6 = 18$

The perimeter of a rectangle is: $2L + 2W$,Perimeter $= 2 \times 18 + 2 \times 6$, Perimeter $= 48$

17) Choice D is correct

The ratio of boy to girls is $7:4$. Therefore, there are 7 boys out of 11 students. To find the answer, first divide the total number of students by 11, then multiply the result by 7.

$55 \div 11 = 5 \Rightarrow 5 \times 7 = 35$. There are 35 boys and 20 $(55 - 35)$ girls. So, 15 more girls should be enrolled to make the ratio $1:1$

18) Choice D is correct

The sum of 8 numbers is greater than 320 and less than 480. Then, the average of the 8 numbers must be greater than 40 and less than 60.

$$\frac{320}{8} < x < \frac{480}{8} \rightarrow 40 < x < 60$$

The only choice that is between 40 and 60 is 45.

19) Choice D is correct

Let x be number of gallons the tank can hold when it is full. Then: $\frac{5}{2}x = 35 \rightarrow x = \frac{2}{5} \times 35 = 14$

20) Choice B is correct

If two triangles are similar, then the ratios of corresponding sides are equal.

$\frac{AC}{AE} = \frac{BC}{DE} = \frac{20}{10} = 2 , \frac{AC}{AE} = 2$

This ratio can be used to find the length of AC: $AC = 2 \times AE$

$AC = 2 \times 10 \rightarrow AC = 20$

The length of AE is given as 10 and we now know the length of AC is 20, therefore:

$EC = AC - AE , EC = 20 - 10, EC = 10$

21) Choice B is correct

In order to solve for the variable b, first take square roots on both sides: $\sqrt{f} = \frac{cz}{b}$, then multiply both sides by b: $b\sqrt{f} = cz$. Now, divide both sides by $\sqrt{f}$: $b = \frac{cz}{\sqrt{f}}$

22) Choice A is correct

Column A: $5^2 = 25$, Column B: $\sqrt[3]{125} = 5$ (recall that $5^3 = 125$)

23) Choice B is correct

A number raised to the exponent $(\frac{1}{2})$ is the same thing as evaluating the square root of the number. Therefore: $(52)^{\frac{1}{2}} = \sqrt{52}$

Since $\sqrt{49}$ is smaller than $\sqrt{52}$, column A $(\sqrt{49} = 7)$ is smaller than $\sqrt{52}$.

24) Choice A is correct

The average is the sum of all terms divided by the number of terms. $21 + 29 + 37 = 87$,

$87 \div 3 = 29$, This is greater than 28.

25) Choice B is correct

Since both columns have 435 as a factor, we can ignore that number.

$16 \times 25 = 400$, $19 \times 22 = 418$, Column B is greater.

26) Choice D is correct

Since x is an integer and can be positive and negative, then the relationship cannot be determined from the information given. Let's choose some values for x.

$x = 1$, then the value in column A is smaller. $-1 < \frac{1}{3}$

Let's choose a negative value for x. $x = -1$, then the value in column A is greater.

$1 > \frac{-1}{3} \rightarrow 1 > -\frac{1}{3}$

27) Choice C is correct

To raise a quantity to a negative power, invert the numerator and denominator, and then raise the base to the indicated power. Therefore:

$(\frac{4}{1})^{-3} = (\frac{1}{4})^3$, The Columns are the same value.

28) Choice A is correct

First, simplify the inequality: $3x + 7 > x - 1 \rightarrow 3x - x > -1 - 7 \rightarrow 2x > -8 \rightarrow x > -4$

29) Choice A is correct

First, find the values of x in both columns.

Column A: $8|3x - 2| = 16 \rightarrow |3x - 2| = 2$

$3x - 2$ can be 2 or -2.

$3x - 2 = 2 \rightarrow 3x = 4 \rightarrow x = \frac{4}{3}$

$3x - 2 = -2 \rightarrow 3x = 0 \rightarrow x = 0$

Column B: $8|3x + 2| = 16 \rightarrow |3x + 2| = 2$

$3x + 2$ can be 2 or -2.

$3x + 2 = 2 \rightarrow 3x = 0 \rightarrow x = 0$

$3x + 2 = -2 \rightarrow 3x = -4 \rightarrow x = -\frac{4}{3}$

The greatest value of x in column A is $\frac{4}{3}$ and the greatest value of x in column B is 0.

30) Choice D is correct

Simplify both columns.

Column A: $(x)^5(x)^2 = x^7$

Column B: $(x^5)^2 = x^{10}$

Column A evaluates to x^7 and Column B evaluates to x^{10}. In the case where $x = 0$, the two columns will be equal, but if $x = 2$, the two columns will not be equal. Consequently, the relationship cannot be determined.

31) Choice D is correct

The probability that an event will occur + the probability that that event will NOT occur must equal 1. Since we don't have any numerical information about the probability, it is possible that the probability that event x occurs is 25%, 50% or any other percent. The probability that event x will not occur will always be 100% minus the probability that event x does occur. Because both columns can exhibit a range of values, the relationship cannot be determined.

32) Choice A is correct

Let x be the original price of the sport jacket. The selling price of a sport jacket including 20% discount is $68. Then: $x - 0.20x = 68 \rightarrow 0.80x = 68 \rightarrow x = \frac{68}{0.80} = 85$

The original price of the jacket is $85 which is greater than column B ($80).

33) Choice D is correct

Factor the expression if possible. Begin by moving all terms to one side before factoring:

$x^2 - 2x - 20 = 15$

$x^2 - 2x - 35 = 0$

To factor this quadratic, find two numbers that multiply to -35 and sum to -2:

$(x - 7)(x + 5) = 0$

Set each expression in parentheses equal to 0 and solve: $x - 7 = 0$

$x = 7$, $x + 5 = 0$, $x = -5$

Quadratic equations can have TWO possible solutions. Since one of these is greater than 5 and one of them is less than 5, we cannot determine the relationship between the columns.

34) Choice B is correct

Recall that numbers between 0 and 1 when raised to power of positive integers become smaller. For example, $(0.5)^2 = 0.25$.

Then: $(0.82)^{27} > (0.82)^{28}$

35) Choice B is correct

Because of the word "and" the events described in each column must be calculated separately and then multiplied: For column A: Probability of rolling a 4: $\frac{1}{6}$

Probability of getting heads: $\frac{1}{2}$, $\frac{1}{6} \times \frac{1}{2} = \frac{1}{12}$

For column B: Probability of an odd number: $\frac{3}{6} = \frac{1}{2}$

Probability of getting a spade: $\frac{13}{52} = \frac{1}{4}$, $\frac{1}{2} \times \frac{1}{4} = \frac{1}{8}$

Since $\frac{1}{8}$ is a larger number than $\frac{1}{12}$, Colum B is greater

36) Choice A is correct

Sum of one quarter, three nickels, and three pennies is: $\$0.25 + 3(\$0.05) + \$0.03 = \$0.43$

37) Choice A is correct

Let's consider the properties of odd and even integers:

$Odd +/- Odd = Even$

$Even +/- Even = Even$

$Odd +/- Even = Odd$

$Odd \times Odd = Odd$

$Even \times Even = Even$

$Odd \times Even = Even$

Now let's review the columns.

For column A: $x(x + y)$

$(odd)(odd + even)$

$(odd)(odd)$

(odd)

For Column B:

$(x - y) - y^2$

$(odd - even) - (even)^2$

$(odd) - (even)(even)$

$(odd) - (even)$

(odd)

Since an odd number is considered greater according to the problem statement, the answer is A.

ISEE Upper Level Math Practice Test 2

Mathematics Achievement

1) Choice D is correct.

Plug in each pair of numbers in the equation. The answer should be 20.

A. $(2, 1)$: $4(2) + 6(1) = 14$ No!

B. $(-1, 3)$: $4(-1) + 6(3) = 14$ No!

C. $(-2, 2)$: $4(-2) + 6(2) = 4$ No!

D. $(2, 2)$: $4(2) + 6(2) = 20$ Yes!

2) Choice D is correct

Let x be the integer. Then: $2x - 5 = 91$, Add 5 both sides: $2x = 96$, Divide both sides by 2: $x = 48$.

3) Choice A is correct

First, multiply both sides of inequality by 5. Then: $\frac{|3+x|}{5} \leq 8 \rightarrow |3 + x| \leq 40$

$-40 \leq 3 + x \leq 40 \rightarrow -40 - 3 \leq x \leq 40 - 3 \rightarrow -43 \leq x \leq 37$

4) Choice B is correct

Subtract $\frac{1}{5b}$ and $\frac{1}{b^2}$ from both sides of the equation. Then: $\frac{1}{5b^2} + \frac{1}{5b} = \frac{1}{b^2} \rightarrow \frac{1}{5b^2} - \frac{1}{b^2} = -\frac{1}{5b}$

Multiply both numerator and denominator of the fraction $\frac{1}{b^2}$ by 5. Then: $\frac{1}{5b^2} - \frac{5}{5b^2} = -\frac{1}{5b}$

Simplify the first side of the equation: $-\frac{4}{5b^2} = -\frac{1}{5b}$

Use cross multiplication method: $20b = 5b^2 \rightarrow 20 = 5b \rightarrow b = 4$

5) Choice A is correct

The sum of supplement angles is 180°. Let x be that angle. Therefore, $x + 9x = 180°$

$10x = 180°$, divide both sides by 10: $x = 18°$

6) Choice B is correct

$x\% \ 26 = 1.3 \rightarrow \frac{x}{100} 26 = 1.3 \rightarrow x = \frac{1.3 \times 100}{26} = 5$

7) Choice B is correct

Plug in the value of $x = 20$ into both equations. Then: $C(x) = x^2 + 2x = (20)^2 + 2(20) = 400 + 40 = 440$, $R(x) = 40x = 40 \times 20 = 800$,$800 - 440 = 360$, So, the profit is \$360.

8) Choice C is correct

$7x^3y^3(2x^3y)^3 = 7x^3y^3(8x^9y^3) = 56x^{12}y^6$

9) Choice D is correct

From choices provided, only choice D is correct. $E = 5 + A, A = S - 4$

10) Choice A is correct

Use Pythagorean Theorem: $a^2 + b^2 = c^2 \Rightarrow 4^2 + 3^2 = c^2 \Rightarrow 25 = c^2 \Rightarrow c = 5\ cm$

11) Choice C is correct

$24\ hours = 86,400\ seconds$, $1,520\ minutes = 91,200\ seconds$

$3\ days = 72\ hours = 259,200\ seconds$

12) Choice C is correct

$C = 2\pi r \Rightarrow C = 2\pi \times 5 = 10\pi \Rightarrow \pi = 3.14 \to C = 10\pi = 31.4$ inches

13) Choice C is correct

Let's review the choices provided:

A. $(6 \times 10^3) + (2 \times 10^2) + (3 \times 10) = 6,000 + 200 + 30 = 6,230$

B. $(6 \times 10^2) + (2 \times 10^1) - 5 = 600 + 20 - 5 = 615$

C. $(6 \times 10^2) + (2 \times 10^1) + 3 = 600 + 20 + 3 = 623$

D. $(6 \times 10^1) + (2 \times 10^2) + 3 = 60 + 200 + 3 = 263$

Only choice C equals to 623.

14) Choice B is correct

First draw an isosceles triangle. Remember that two sides of the triangle are equal.

Let put a for the legs. Then: Use Pythagorean theorem to find the value of a. isosceles right triangle

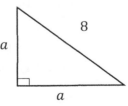

$a^2 + b^2 = c^2 \to a^2 + a^2 = 8^2$

Simplify: $2a^2 = 64 \to a^2 = 32 \to a = \sqrt{32}$

$a = \sqrt{32} \Rightarrow$ area of the triangle is $= \frac{1}{2}(\sqrt{32} \times \sqrt{32}) = \frac{1}{2} \times 32 = 16\ cm^2$

15) Choice C is correct

$\frac{5}{500} = \frac{x}{860} \Rightarrow x = \frac{5 \times 860}{500} = \$\ 8.6$

16) Choice B is correct

Area of a circle equals: $A = \pi r^2$, The new diameter is 20% larger than the original then the new radius is also 20% larger than the original. 20% larger than r is $1.2r$. Then, the area of larger circle is: $A = \pi r^2 = \pi(1.2r)^2 = \pi(1.44r^2) = 1.44\pi r^2$. $1.44\pi r^2$ is 44% bigger than πr^2.

17) Choice B is correct

$89.44 \div 0.05 = 1,788.8$

18) Choice A is correct

The area of the non-shaded region is equal to the area of the bigger rectangle subtracted by the area of smaller rectangle. Area of the bigger rectangle $= 14 \times 18 = 252$

Area of the smaller rectangle $= 9 \times 3 = 27$, Area of the non-shaded region $= 252 - 27 = 225$

19) Choice D is correct

$2\frac{1}{2} - 1\frac{5}{4} =$, Break off 1 from 2: $2\frac{1}{2} = 1\frac{3}{2}$

$1\frac{3}{2} - 1\frac{5}{4} =$ Subtract whole numbers: $1 - 1 = 0$, Combine fractions: $\frac{3}{2} - \frac{5}{4} = \frac{1}{4}$

20) Choice D is correct

To find the maximum value of y, the expression $(x - 2)^2$ must be equal to 0. Because it has a negative sign. Since $x - 2$ is to the power of 2, it cannot be negative. To get 0 for the expression $(x - 2)^2$, x must be 2. Plug in 2 for x in the equation: $y = -(x - 2)^2 + 7 \rightarrow y = -(2 - 2)^2 + 7 = 7$, The maximum value of y is 7.

21) Choice B is correct

$\begin{cases} -3x - y = -5 \\ 5x - 5y = 15 \end{cases} \Rightarrow$ Multiplication (-5) in first equation $\Rightarrow \begin{cases} 15 + 5y = 25 \\ 5x - 5y = 15 \end{cases}$

Add two equations together $\Rightarrow 20x = 40 \Rightarrow x = 2$ then: $y = -1$

22) Choice C is correct

Let's review the choices provided. Put the values of x and y in the equation.

A. $(2, 7)$ $\Rightarrow x = 1 \Rightarrow y = 7$ This is true!

B. $(-2, -13)$ $\Rightarrow x = -2 \Rightarrow y = -13$ This is true!

C. $(4, 21)$ $\Rightarrow x = 4 \Rightarrow y = 17$ This is not true!

D. $(-4, -23)$ $\Rightarrow x = 2 - 4 \Rightarrow y = -23$ This is true!

23) Choice C is correct

To find total number of miles driven by Ed that week, you only need to subtract 53,431 from 52,806. $53,431 - 52,806 = 625 \ miles$

24) Choice A is correct

Isosceles right triangle

First draw an isosceles triangle. Remember that two sides of the triangle are equal.

Let put a for the legs. Then:

$a = 4 \Rightarrow$ area of the triangle is $= \frac{1}{2}(4 \times 4) = \frac{16}{2} = 8 \ cm^2$

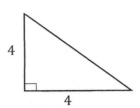

25) Choice A is correct

Factor each trinomial $x^2 - 5x + 6$ and $x^2 - 6x + 8$

$x^2 - 5x + 6 \Rightarrow (x-2)(x-3)$, $x^2 - 6x + 8 \Rightarrow (x-2)(x-4)$

The common factor of both expressions is $(x - 2)$.

26) Choice C is correct

$$\begin{array}{r} 36 \text{ hr. } 38 \text{ min.} \\ - 23 \text{ hr. } 25 \text{ min.} \\ \hline 13 \text{ hr. } 13 \text{min} \end{array}$$

27) Choice B is correct

$\frac{14}{26} = 0.538 \cong 0.54$

28) Choice C is correct

$x + y = 10$, Then: $9x + 9y = 9(x + y) = 9 \times 10 = 90$

29) Choice C is correct

First, convert all measurement to foot. One foot is 12 inches. Then: 12 inches = $\frac{10}{12} = \frac{5}{6}$ feet

The volume flower box is: length × width × height $= 2 \times \frac{5}{6} \times 2 = \frac{10}{3}$ cubic feet.

30) Choice A is correct

$\frac{460}{20} = 23$ miles per gallon

31) Choice D is correct

$\frac{x^3}{15} \Rightarrow$ reciprocal is : $\frac{15}{x^3}$

32) Choice B is correct

Let's write equations based on the information provided:

$Michelle = Karen - 9$

$Michelle = David - 4$

$Karen \ + \ Michelle \ + \ David \ = \ 91$

$Karen - 9 \ = \ Michelle \Rightarrow Karen \ = \ Michelle + 9$

$Karen \ + \ Michelle \ + \ David \ = \ 91$

Now, replace the ages of Karen and David by Michelle. Then:

$Michelle + 9 \ + \ Michelle \ + \ Michelle + 4 \ = \ 91$

$3Michelle + 13 = 91 \Rightarrow \ 3Michelle = 91 - 13$

$3Michelle = 78$

$Michelle = 26$

33) Choice C is correct

Use interest rate formula: $Interest = principal \times rate \times time = 1{,}400 \times 0.06 \times 1 = \$84$

34) Choice D is correct

$A = bh$, $A = 3 \times 3.2 = 9.6$ square feet

35) Choice D is correct

Ellis travels $\frac{5}{3}$ of 90 hours. $\frac{5}{3} \times 90 = 150$, Ellis will be on the road for 150 hours.

36) Choice A is correct

Plug in the values of x and y in the expression:

$2x^2(y \ + \ 4) \ = \ 2(0.5)^2(5 \ + \ 4) \ = \ 2 \ (0.25)(9) \ = \ 4.5$

37) Choice B is correct.

To find the area of the shaded region subtract smaller circle from bigger circle.

$S_{bigger} - S_{smaller} = \pi \left(r_{bigger}\right)^2 - \pi \ (r_{smaller})^2 \Rightarrow S_{bigger} \ - \ S_{smaller} = \pi(7)^2 - \pi \ (5)^2 \Rightarrow 49 \ \pi - 25\pi \ = \ 24 \ \pi \ inch^2$

38) Choice D is correct

To find the discount, multiply the number by $(100\% - rate \ of \ discount)$.

Therefore, for the first discount we get: $(500) \ (100\% - 25\%) = (500) \ (0.75)$

For the next 15% discount: $(500) \ (0.75) \ (0.85)$

39) Choice A is correct

Write a proportion and solve for the missing number. $\frac{40}{18} = \frac{5}{x} \rightarrow 40x = 18 \times 5 = 90$

$$40x = 90 \rightarrow x = \frac{90}{40} = 2.25 \, ft$$

40) Choice A is correct

The ratio of boys to girls is $7:3$. Therefore, there are 7 boys out of 10 students. To find the answer, first divide the number of boys by 7, then multiply the result by 10.

$$210 \div 7 = 30 \Rightarrow 30 \times 10 = 300$$

41) Choice B is correct

$$\frac{35}{100} \times 620 = x \rightarrow x = 217$$

42) Choice C is correct

The area of a $19 \, feet \times 19 \, feet$ room is 361 square feet. $19 \times 19 = 361$

43) Choice C is correct

Use FOIL (First, Out, In, Last).$(4x + 4)(x + 5) = 4x^2 + 20x + 4x + 20 = 4x^2 + 24x + 20$

44) Choice C is correct

Plug in the values of x and y in the equation: $4\blacksquare9 = \sqrt{4^2 + 9} = \sqrt{16 + 9} = \sqrt{25} = 5$

45) Choice A is correct

Let x be the capacity of one tank. Then, $\frac{2}{3}x = 200 \rightarrow x = \frac{200 \times 3}{2} = 300$ Liters

The amount of water in four tanks is equal to: $4 \times 300 = 1{,}200$ Liters

46) Choice C is correct.

To add two matrices, first we need to find corresponding members from each matrix.

$$\begin{vmatrix} 3 & 6 \\ -1 & -3 \\ -5 & -1 \end{vmatrix} + \begin{vmatrix} 2 & -1 \\ 6 & 4 \\ 1 & 3 \end{vmatrix} = \begin{vmatrix} 5 & 5 \\ 5 & 1 \\ -4 & 2 \end{vmatrix}$$

47) Choice C is correct

$$Average = \frac{\text{sum of terms}}{\text{number of terms}}$$

The sum of the weight of all girls is: $20 \times 55 = 1{,}100 \, kg$, The sum of the weight of all boys is: $35 \times 70 = 2{,}450 \, kg$, The sum of the weight of all students is: $1{,}100 + 2{,}450 = 3{,}550 \, kg$

$$Average = \frac{3{,}550}{55} = 64.54 \, kg$$

"Effortless Math Education" Publications

Effortless Math authors' team strives to prepare and publish the best quality ISEE Upper Level Mathematics learning resources to make learning Math easier for all. We hope that our publications help you learn Math in an effective way and prepare for the ISEE Upper Level test.

We all in Effortless Math wish you good luck and successful studies!

Effortless Math Authors

www.EffortlessMath.com

… So Much More Online!

- ❖ FREE Math lessons

- ❖ More Math learning books!

- ❖ Mathematics Worksheets

- ❖ Online Math Tutors

Need a PDF version of this book?

Visit www.EffortlessMath.com

Receive the PDF version of this book or get another FREE book!

Thank you for using our Book!

Do you LOVE this book?

Then, you can get the PDF version of this book or another book absolutely FREE!

Please email us at:

info@EffortlessMath.com

for details.